Mere Thermodynamics

Mere Thermodynamics

DON S. LEMONS

The Johns Hopkins University Press

Baltimore

© 2009 The Johns Hopkins University Press
All rights reserved. Published 2009
Printed in the United States of America on acid-free paper
9 8 7 6 5 4 3 2 1

The Johns Hopkins University Press
2715 North Charles Street
Baltimore, Maryland 21218-4363
www.press.jhu.edu

Library of Congress Cataloging-in-Publication Data

Lemons, Don S. (Don Stephen), 1949–
    Mere thermodynamics / Don S. Lemons.
        p. cm.
    Includes bibliographical references and index.
    ISBN-13: 978-0-8018-9014-7 (hardcover : alk. paper)
    ISBN-10: 0-8018-9014-4 (hardcover : alk. paper)
    ISBN-13: 978-0-8018-9015-4 (pbk. : alk. paper)
    ISBN-10: 0-8018-9015-2 (pbk. : alk. paper)
    1. Thermodynamics. I. Title.
    QC311.L358 2008
    536'.7—dc22   2008007841

A catalog record for this book is available from the British Library.

*Special discounts are available for bulk purchases of this book. For more information,
please contact Special Sales at 410-516-6936 or specialsales@press.jhu.edu.*

The Johns Hopkins University Press uses environmentally friendly book materials,
including recycled text paper that is composed of at least 30 percent post-consumer
waste, whenever possible. All of our book papers are acid-free, and our jackets and
covers are printed on paper with recycled content.

Q2

Contents

....

Thermodynamics is a physical theory at once beautiful and useful: beautiful because its laws are simply expressed and deductions from them universally applicable, and useful because in distinguishing the possible from the impossible thermodynamics saves us from much fruitless effort. For these reasons, as well as for its undeniable empirical success, thermodynamics deeply impresses those who grasp its essentials. Albert Einstein once boldly claimed that thermodynamics, alone among physical theories, "will never be overthrown."

Thermodynamics is nevertheless notoriously difficult to teach and difficult to learn. While much of classical and quantum physics can be framed in terms of easily pictured concepts or easily remembered equations, classical thermodynamics grows out of the austere logic of possibility and impossibility encapsulated in its first and second laws. That these laws are usually given verbal rather than mathematical expression hinders rather than aids readers unused to such formulations.

Even so, most texts don't bother much with the laws of thermodynamics. While no one questions the laws per se, everyone hurries to get beyond them. What lies beyond the laws of thermodynamics is, on the one hand, a multitude of applications and, on the other, the special models of classical and quantum statistical mechanics. The diligent student of such applications and such models, while able to reproduce many complicated patterns of thought, may lack a coherent vision of thermodynamics itself. On too many occasions accomplished scientists have told me, "I never understood thermodynamics."

Mere Thermodynamics presents a vision of thermodynamics itself, its laws, their essential corollaries, useful methods, and important applications. This ordering—first laws, then corollaries, methods, and applications—informs the whole text. My aim has been to present the subject in its most orderly and most plausible aspect, and to write a concise yet balanced text that allows the structure of classical thermodynamics to stand out—its limitations as well as its achievements.

In highlighting the intellectual structure of thermodynamics I have found it impossible to ignore the historical drama of its unfolding. Certain contours of the subject have been permanently shaped by its history, and for this reason, this volume follows the traditional historically oriented approach with its heat engines, reversible cycles, and laws. The historical approach also naturally introduces, in its course, important phenomena and experimental outcomes.

The text aims to reward the reader's attention with a maximal understanding of the subject's most difficult parts: the second law of thermodynamics and the concept of entropy. Formally, the meaning of any statement or law consists of what can be deduced from it. For this reason, and also because the second law was discovered before the first, I develop consequences of each law apart from the other—the first law in Chapter 4 and the second law in Chapter 5 and Appendix B—before developing, in Chapter 6, consequences of the two combined. The first and second laws of thermodynamics together lead to the concept of entropy. There is hardly any development of physical theory more impressive than that taking us from the first law and a simple verbal statement of the second law to the existence of entropy as a state variable. This transition, from words to mathematical expression, in Chapter 7, is crucial to thermodynamics. But it also illustrates the distinction between theoretical physics, which in its fullness continually makes transitions of this sort, and the mathematical manipulation of physically meaningful variables.

Chapters 1–7 present concepts, laws, and important corollaries, Chapters 8–9 useful methods and essential applications. This material and related end-of-chapter problems compose a brief course on classical thermodynamics. The remaining four chapters, 10–13, present more loosely sequenced, if fairly standard, topics: nonfluid variables, equilibrium and stability, two-phase systems, and the third law. In spite of the book's title, I do, on occasion, invoke the molecular hypothesis—that matter is composed of molecules—in order to motivate unfamiliar equations of state. But statistical methods, quantum concepts, and chemical reactions define, by their exclusion, the boundary of *Mere Thermodynamics*.

This book is designed for second-, third-, and fourth-year physics, chemistry, and pre-engineering undergraduate students who have studied or are concurrently studying multivariate calculus. Often these students have been inadequately introduced to basic thermodynamics in their first physics course. Sometimes, in my own course, I create time for students to report on topics that address their own interests in thermodynamics.

The Annotated Bibliography describes several books and a few articles that provide foundation for, complement, or build upon the ideas presented here. I am indebted to their authors. But I am especially pleased to acknowledge students, friends, and colleagues who have personally contributed to *Mere Thermodynamics*. Ralph Baierlein, Galen Gisler, and Joel Krehbiel each read and commented on the whole text; Jeff Buller, Rickey Faehl, Bob Harrington, Blake Johnson, Rhon Keinigs, Dwight Neunschwander, Margaret Penner, Bill Peter, Paul Regier, and Bryce Schmidt each read and commented on part of the text. Chapter 13, on the third law of thermodynamics, could not have been written without the aid of Ralph Baierlein, who gave generously of his time and expertise. Margaret Penner's senior thesis inspired Appendix B, on the logical consequences of the second law as shown in 21 simple cycles. Willis Overholt skillfully drew or redrew the book's figures. I wrote much of the text during a sabbatical leave from Bethel College of North Newton, Kansas. Finally, I offer heartfelt thanks to my editor at the Johns Hopkins University Press, Trevor Lipscombe. His encouragement and expertise helped me enjoy the writing process. I dedicate this book to my wife, Allison, and two sons, Nathan and Micah. They stood by me during difficult times.

Mere Thermodynamics

Definitions

1.1 Thermodynamics

The Greek roots of the word *thermodynamics*, *thermo* (heat) and *dynamics* (power or capacity), neatly compose a definition. Etymologically, thermodynamics means the power created by heat, or as we would now say, the work created by heat. Because engines of many kinds produce work from heat, their study belongs to the science of thermodynamics. The concept of work is familiar from mechanics, but what is heat? Thermodynamics assigns its own special meanings not only to the word *heat* but also to the terms *system*, *boundary*, and *state*.

1.2 System

A thermodynamic *system* is simply that part of the universe with which we are concerned. We may, for instance, focus on a bucket of seawater or on a beam of iron. Thermodynamic systems may be composed of several chemically distinct components (like seawater), or exist in several phases (as solid, liquid, or gas), or occupy spatially separate parts. We make progress most rapidly by attending first to simple homogeneous, single-phase systems.

1.3 Boundary, Environment, and Interactions

Each thermodynamic system is surrounded by a *boundary* separating the system from its *environment*. Boundaries regulate the *interaction* between the system and its environnment or between two systems. Boundaries can be divided into several kinds: those that permit or forbid work to be done on or by the system and those that permit or forbid heat to be absorbed or rejected by the system. For instance, a movable boundary allows mechanical work to be done on or by the system (see Fig. 1.1), while a rigid one does not.

1.4 States and State Variables

A set of quantities called *state variables* define the *state* of each thermodynamic system. State variables include those appropriate to a simple fluid (pressure and volume), to a surface (surface tension

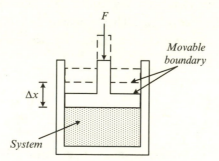

FIGURE 1.1 Idealized thermodynamic system with a movable boundary that allows work $F\Delta x$ to be done on the system.

and area), to black-body radiation (energy density and radiation pressure), and to an electrical contact (contact potential and current). Different sets of state variables conveniently describe different thermodynamic systems. When a system interacts with its environment, its state variables change.

Appropriate state variables can be identified only after a thorough investigation of the system. Thermodynamic state variables are measured with laboratory instruments—that is, macroscopic devices (pressure gauge, balance, and meter stick)—or are inferred from such measurements.

1.5 Equations of State

State variables enter into relations among themselves called *equations of state*. These relations can be defined by tables of data, by numerical fits to those data, or by analytic models. An example of an analytic equation of state, which we take up in Chapter 9, is $P = E/3V$, where P is the pressure, E the internal energy, and V the volume of so-called cavity radiation. The equation of state $V = V_o(1 + \beta_o T - \kappa_o P)$ relates the state variables of a solid, where T is the system temperature and V_o, β_o, and κ_o are characterizing constants. Each equation of state reduces the number of independent state variables by one.

Thermodynamics is concerned only with systems described by state variables that are related by equations of state. A thermodynamic description per se makes no direct claim about a system's ultimate components, about its atoms and molecules, their interactions, and their positions and velocities. Thermodynamics differs from most other sciences in not being reductionist. (See Problem 1.1.)

1.6 Work

Performing work on a system changes its state. Work may be performed in a number of ways and not only, as in Figures 1.1 or

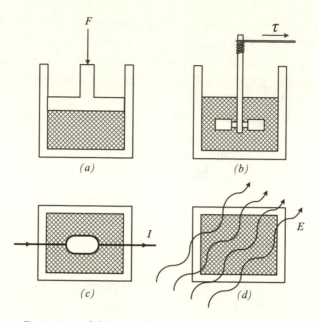

FIGURE 1.2   Four ways of doing work on a system: (a) compressing the system, (b) rotating a paddle wheel immersed in the system, (c) energizing an electric circuit element that is part of the system, and (d) applying an electric field to charges within the system.

1.2a, by pushing a piston into a cylinder that contains the system. Mechanics and electrodynamics books describe different ways of performing work. For instance, a torque τ applied to a rod that turns a paddle through an angle $\Delta\theta$ does work $\tau\Delta\theta$ on the liquid in which the paddle is immersed (Fig. 1.2b). One means of doing electrical work is to include within the system a circuit element across which a potential difference $\Delta\Phi$ is applied. A charge q, in passing through the circuit element, loses energy $q\Delta\Phi$ to the system at a rate $I\Delta\Phi$, where $I = dq/dt$ (Fig. 1.2c). Another way is to apply an externally generated electric field E to the system. Then, as the system charges move and develop an electric dipole moment ΔP, the environment does a quantity of work $E \cdot \Delta P$ on the system (Fig. 1.2d). (See Problem 1.2.)

A system may interact with its environment in ways other than work. We know this because on occasion the state of a system changes even when no work is done on the system. One has only to think of what happens to a cup of hot coffee sitting on a table; inevitably it becomes colder. Let's imagine a system with idealized boundary. The boundary is rigid, prohibiting mechanical work interactions; metallic, through which an externally applied electric field cannot penetrate; and impermeable, so that mass and charge cannot enter or leave. Further imagine that this idealized boundary prohibits every other kind of work interaction. Even so, another kind of interaction with the environment can still change the system's state variables. By definition, that which causes the state of a system to change during a nonwork interaction through an impermeable boundary is *heat*.

Of course, we can also conceive of an *adiabatic* boundary that, by definition, prohibits heat interactions, that is, heating and cooling. There are practical ways of constructing an adiabatic boundary. A so-called Dewar flask, named after its inventor Sir James Dewar (1842–1923), with rigid, vacuum-containing, and radiation-reflecting surfaces approximates an adiabatic, work-prohibiting boundary.

A boundary that allows heating is called *diathermal*. Systems that interact through a diathermal boundary are said to be in *thermal contact*. Figure 1.3 illustrates various combinations of boundaries that allow only heat, only work, both heat and work, and neither heat nor work interactions. An adiabatic, work-prohibiting boundary that is also closed to mass transfer completely *isolates* a system from its environment. (See Problem 1.3.)

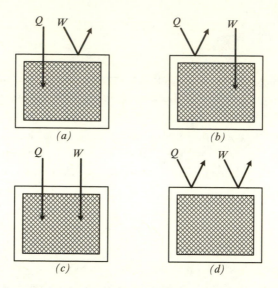

FIGURE 1.3 Boundaries that prohibit and allow work and heat interactions: (a) diathermal, work-prohibiting; (b) adiabatic, work-allowing; (c) diathermal, work-allowing; (d) adiabatic and work-prohibiting.

Chapter 1 Problems

1.1 *Definitions.* Define *system, boundary, adiabatic boundary, diathermal boundary, environment, heat, state variables,* and *equation of state* with a phrase or sentence. Be concise and be complete.

1.2 *Work or Heat?* You grab a bottle of juice and shake it thoroughly. Is this an example of a work or a heat interaction? Recall that work can be described with mechanical or electromagnetic variables.

1.3 *Interactions.* In each of the following interactions indicate whether the system does work, has work done on it, or does no work and whether the system boundary is diathermal or adiabatic.

(a) The system is the air contained within a bicycle tire along with a tire pump connected to it. The pump plunger is pushed down, forcing air into the tire. Assume this interaction is over before the air significantly cools. (S)[1]

(b) The system is the water and water vapor within a metal pot covered by a tight-fitting lid. The pot is placed over a flame. The temperature and pressure of the water and water vapor increase.

(c) Following the interaction in part (b) the lid flies off the pot. Assume this happens very quickly.

(d) The system is the air contained within a room. The room doors and windows are closed and are of ordinary construction. A fan within the room is left running for one hour.

(e) The system is a mixture of hydrogen and oxygen. This combustible mixture is contained within a rigid chamber. The mixture explodes.

---

[1] "S" here and elsewhere indicates that a complete solution is found in Appendix D.

TWO

Equilibrium

....

2.1 Equilibrium

....

Classical thermodynamics is concerned with *equilibrium states* described by a small set of macroscopic variables that change only when the system's environment changes. Relatively sudden or violent interactions, such as those caused by pouring or stirring a fluid, destroy equilibrium because they set in motion changes that persist even after the interaction is complete. Given time, an isolated system not in equilibrium will evolve toward an equilibrium state that, when established, will persist indefinitely.

Two systems are in *mutual* equilibrium if the state variables of neither change when the two interact. One can further distinguish between several kinds of mutual equilibrium: one for each kind of interaction possible between systems, for example, thermal, mechanical, chemical, electrical, and magnetic.

The concept of *thermal* equilibrium is unique to the science of thermodynamics. Two systems are in mutual thermal equilibrium if the state variables of neither change when the two are placed in thermal contact. When two initially isolated systems, not in mutual thermal equilibrium, are placed in thermal contact, they eventually achieve mutual thermal equilibrium.

2.2 Zeroth Law of Thermodynamics

Experience teaches us that two systems in mutual thermal equilibrium with a third are in thermal equilibrium with each other. Such relations are illustrated in Figure 2.1. Apparently, the relation of thermal equilibrium propagates through a third system interposed between two others as long as each is in thermal contact with its immediate neighbors. In this way the relation of thermal equilibrium is like the equality relation of arithmetic: both are transitive in that they carry across a middle term. Relatively late in the history of thermodynamics R. H. Fowler (1899–1944) elevated this universally observed fact to the status of a law: *the zeroth law of thermodynamics.*

The zeroth law of thermodynamics allows us to specify a single thermodynamic system as an indicator of thermal equilibrium. Suppose this specially chosen equilibrium-indicating system or *thermometer* is brought successively into thermal contact with two systems and allowed to achieve equilibrium with each. If the thermometer's state variables do not change when brought into successive contact

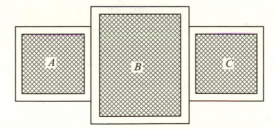

FIGURE 2.1   The zeroth law of thermodynamics: two systems, A and C, in thermal equilibrium with a third, B, are in thermal equilibrium with each other.

with the two systems, the two systems are in thermal equilibrium. If the thermometer state variables do change, the two systems are not in thermal equilibrium.

2.3 Empirical Temperature

The zeroth law of thermodynamics allows us to associate with each equilibrium state a state variable T, called the temperature, chosen in accordance with the rule that any two systems in thermal equilibrium have the same temperature and any two systems not in thermal equilibrium have different temperatures. So chosen, a system's temperature reveals all its relations of potential thermal equilibrium or disequilibrium with other systems whose temperature is known.

While the zeroth law itself does not dictate a method of assigning temperatures to equilibrium states, scientists naturally prefer the convenience of a standard method. Recall that a thermometer brought into successive thermal contact with two systems reveals whether or not the two systems are or could be in thermal equilibrium. Why not designate one variable of one indicator, the other variables remaining fixed, as the temperature-measuring—that is, *thermometric*—variable of the thermometer? Temperatures based in this way on the equilibrium states of a particular thermometric system are called *empirical temperatures*.

One example of a thermometric property is the volume of the water-alcohol mixture in the common outdoor thermometer. The liquid is often colored red for easy viewing. How far the liquid rises in its small-bore glass tube represents its volume. Marks on the glass tube, corresponding to different volumes of the liquid, are denominated in degrees of temperature according to a rule or *temperature scale.*

Other kinds of thermometers are, of course, possible and have their uses. Environments that would freeze or vaporize the water-alcohol mixture or that allow no room for the thermometer bulb to be proximately placed render the water-alcohol thermometer impractical. In these cases a resistance thermometer with digital display might be appropriate. Resistance thermometers use the electrical resistance of a piece of metal, often platinum, as the thermometric property. When the platinum strip is brought into thermal equilibrium with another object, its resistance, calibrated in degrees, indicates the object's temperature. Optical pyrometers employ the characteristic color of radiation emitted by hot objects as a thermometric property. According to the zeroth law several thermometers may be used simultaneously if each is calibrated to return the same temperature when brought into mutual thermal equilibrium with the same system. (See Problem 2.1.)

2.4 Traditional Temperature Scales

The choice of temperature scale is distinct from the choice of thermometric variable. All useful scales, however, require the temperature, T, to be an ever-increasing or ever-decreasing function $T(X)$ of the thermometric variable X. If the scale were not monotonic, two systems not in thermal equilibrium might be assigned the same temperature.

Traditional *two-point temperature scales* are defined by a linear function $T(X) = aX + b$, where X is the thermometric variable of the standard thermometer and a and b are constants that characterize the scale. The characterizing constants a and b are chosen in order that values of the thermometric variable at two sensitive but easily reproduced *fixed points*—that is, X_1 and X_2—correspond to stipulated temperatures T_1 and T_2. Typical fixed points are the melting point of ice at atmospheric pressure—that is, the *normal ice point*—and the boiling point of liquid water at atmospheric pressure—that is, the *normal steam point*. For instance, the Celsius scale is defined by assigning a temperature of 0°C to the normal ice point and 100°C to the normal steam point of water. The Fahrenheit scale assigns 32°F to the normal ice point and 212°F to the normal steam point of water. (See Problems 2.2–2.3.)

2.5 Equilibrium Processes

Changes or *processes* that unfold very slowly—so slowly that the system always remains in, or arbitrarily close to, equilibrium even as it passes from one state to another—are called *equilibrium* or *quasistatic processes*. Quasistatic processes can be analyzed into indefinitely small, temporally ordered parts, each standing for an equilibrium state, and can be represented by continuous lines on a *state variable diagram*. Of course, quasistatic processes are idealizations; they can be approximated but never fully realized.

Figure 2.2 illustrates a quasistatic process generated by a *Joule apparatus*—a paddle wheel inserted into a constant-volume fluid system contained within an adiabatic boundary. The handle of the Joule apparatus turns the paddles very slowly, that is, quasistatically, and very slowly causes the temperature of the fluid to increase. The state variable diagram of Figure 2.2b represents the sequence of continuously spaced equilibrium states through which the fluid moves as the handle turns.

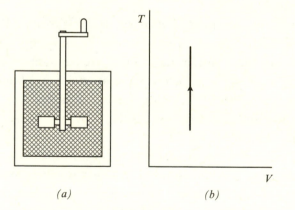

(a)   *(b)*

FIGURE 2.2   A paddle wheel quasistatically stirs a constant-volume system contained within an adiabatic boundary: (a) Joule apparatus with handle; (b) state variable diagram representing the sequence of equilibrium states through which the system passes.

Chapter 2 Problems

2.1. *Resistance Thermometer.* The resistance of a platinum wire is found to be 7.000 ohms at the normal ice point of water (0.000°C and 1 atm), 9.705 ohms at the normal steam point of water (100.0°C and 1 atm), and 18.39 ohms at the normal melting point of sulfur (444.1°C and 1 atm). Suppose the resistance, R, versus temperature, T, in degrees Celsius falls close to the curve

$$R(T) = R_o \left(1 + aT + bT^2\right)$$

when R_o, a, and b are determined by measurements at the normal ice, steam, and sulfur points.

(a) Find R_o, a, and b.
(b) Invert $R(T)$ to find the temperature T as a function of R. Require that $T(R)$ be an ever-increasing function of R between the normal ice point and the melting point of sulfur.
(c) Suppose this thermometer is calibrated only at the ice and steam points and the constant b is arbitrarily chosen so that $b = 0$. What temperature would this thermometer calibrated in this way produce for the sulfur point (where $R = 18.39\Omega$)?

2.2. *Traditional Temperature Scales.* At what temperature would adjacent Celsius and Fahrenheit thermometers read the same number?

2.3. *Réaumur Scale.* Russian novels set in the 19th century sometimes refer to the two-point Réaumur temperature scale. According to this scale water freezes at 0 Ré and boils at 80 Ré. Find the Réaumur temperature that corresponds to 90°F.

Heat

....

3.1 Quantifying Heat

....

The concepts of temperature and heat were often confused before Joseph Black (1728–1799) carefully distinguished between the two in the late eighteenth century. While both are uniquely thermodynamic concepts, each plays its own role within the subject. Temperature, for instance, is an *intensive* state variable—intensive because temperature does not depend in a direct way upon the size of the system. After all, two different-sized systems in mutual thermal equilibrium have the same temperature.

In contrast, heat is not a state variable at all but, rather, quantifies the interaction between a system and its environment allowed by a diathermal boundary. Early in the nineteenth century scientists adopted a water-standard definition of the quantity of heat. According to this definition, the absorption of one *calorie* (1 cal) of heat raises the temperature of one gram of water one degree Celsius under standard conditions. (The nutritional calorie, or kilocalorie, always abbreviated with an initial upper-case letter as in Cal, is by convention 1000 calories.) Here "standard conditions" means at 14.5°C and atmospheric pressure. That a gram of water under standard conditions expands, very slightly, as its temperature increases 1°C and in so doing pushes against and does work on the atmosphere in no way diminishes the precision of this operational definition.

In principle, the water-standard definition of a calorie allows one to quantify the heat exchanged in an arbitrary process. One simply arranges for the quantity of heat in question to be absorbed by a quantity of water whose temperature will, under standard conditions, increase by exactly 1°C. That mass of water in grams is the number of calories transferred.

3.2 Calorimetry

Adopting the water-standard definition of the calorie is equivalent to defining the *heat capacity* of one gram of water under standard conditions to be 1.00 cal/g °C. The heat capacity C of any object is the ratio of the quantity of heat dQ absorbed that increases its temperature dT under given conditions. Thus,

$$C = \frac{dQ}{dT}.$$ (3.1)

Heat capacity is an *extensive* state variable that is directly proportional to the size of the system. In contrast, the *specific heat,*

$$c_m = \frac{1}{m}\frac{dQ}{dT},$$ (3.2)

that is, the heat capacity per unit mass, is an intensive variable that characterizes the type of system. Molar specific heats,

$$c_{mol} = \frac{1}{n}\frac{dQ}{dT},$$ (3.3)

TABLE 3.1 Specific heats of common materials at 25°C and atmospheric pressure

Material	Specific heat (cal/g °C)
Aluminum	0.215
Copper	0.093
Gold	0.031
Iron	0.108
Lead	0.031
Mercury	0.033
Silver	0.056
Sodium chloride	0.210
Wood	0.406
Water	1.000

where n is the number of moles in the system, are also useful intensive state variables. The specific heat of aluminum, for instance, is 0.215 cal/g °C. Table 3.1 lists a few specific heats. Note that the specific heats of metals are small compared to that of water.

Heating a system usually increases its temperature, while cooling usually decreases its temperature. But under certain conditions the transfer of heat to or from a system leaves its temperature unchanged. For instance, one gram of boiling water at atmospheric pressure can absorb 539 cal of heat without increasing its temperature. Likewise, one gram of freezing water can reject 80 cal without changing its temperature. Heat transfers that leave a system's temperature unchanged, traditionally called *latent heats*, are associated with phase transitions. The *latent heat of fusion* is the heat necessary to melt one gram of solid at its normal melting point, and the *latent heat of vaporization* is the heat necessary to vaporize one gram of liquid at its normal boiling point. Table 3.2 lists *heats of transition*, normal melting, and normal boiling points of common materials.

Knowing the specific and latent heats of a variety of materials under a range of conditions allows one to determine the heat required to bring two systems not initially in equilibrium into mutual

TABLE 3.2 Latent heats of fusion and vaporization and the melting and boiling points for common materials at atmospheric pressure

Substance	Melting point (°C)	Heat of fusion (cal/g)	Boiling point (°C)	Heat of vaporization (cal/g)
Aluminum	660	95.3	2467	2940
Copper	1083	48.9	2567	1147
Gold	1063	48.9	2660	377
Iron	1535	65.6	2750	1503
Lead	328	5.85	1740	208
Mercury	−39	2.82	357	69.3
Nitrogen	−210	6.10	−196	47.8
Oxygen	−219	3.30	−183	50.2
Water	0	80.0	100	539

thermal equilibrium. The methods of such inference compose the science of *calorimetry*. Suppose, for instance, a 50-g block of 75°C aluminum cools and achieves equilibrium with its 25°C environment. The heat lost by the aluminum during this process is 50 g × (75–25)°C × (0.215 cal/g °C), or 538 cal, provided the specific heat of aluminum remains approximately constant as it cools from 75 to 25°C. Indirect, calorimetric measurements of heat are usually more convenient than ones resorting directly to the water-standard definition of a calorie. (See Problems 3.1–3.4.)

3.3 What is Heat?

If you are still inclined to ask the question "What, exactly, is heat?" you are probably seeking an answer in terms of nonthermodynamic, microscopic quantities. Before Joule's experiments of the 1840s two such answers were possible: (1) heat is its own kind of rarefied, ungenerate, indestructible, and probably massless fluid; and (2) heat is a kind of motion. Either of these answers was consistent with the known calorimetric facts, but both were speculations that went beyond those facts. More importantly, each view is consistent with the definition of heat offered in Chapter 1: that which when transferred to or from a system through a work-prohibiting boundary changes the system's state.

However, the idea that heat is a substance called caloric that is conserved in all processes was, in the early 1800s, the most convincingly straightforward of the two speculations. In thermal interactions the caloric flowed from one body to another; in melting and vaporization caloric was absorbed and stored; and in freezing and condensing caloric was liberated. While the caloric theory of heat is plausible and to this day remains useful in limited circumstances, it ultimately proved inadequate as a foundational theory.

Today we recognize that neither idea—heat as caloric and heat as motion—is completely adequate. Rather, as we will find in Chapter 4, heat is a quantifiable nonwork energy transfer.

Chapter 3 Problems

3.1 *Calorimetry*. How many calories of heat must be transferred to 12 g of copper in order to raise its temperature from 10°C to 35°C?

3.2 *Specific Heat*. One kilogram of hot (80.0°C) aluminum is placed into 10.0 L of 20.0°C water in a container with negligible heat capacity and adiabatic walls. After the aluminum and water have reached thermal equilibrium their common temperature is 21.3°C. Assume the specific heats of water and aluminum are constants. What is the specific heat of aluminum implied by this data?

3.3 *Average Person*. An average person weighs 65 kg and consumes 2000 nutritional calories per day. If these nutritional calories were supplied to 65 kg of water, by how many degrees Celsius would the water increase? How many degrees Fahrenheit?

3.4 *Latent Heat*. How much heat is required to raise the temperature of 50.0 g of H_2O ice at 0.00°C to 30.0°C? Assume an average 1.00 cal/g °C specific heat for water in this temperature range.

The First Law

....

4.1 Count Rumford
....

Benjamin Thompson's various occupations—spy, social reformer, arms manufacturer, and inventor—never kept him from aggressively promoting his own cause. He sought high office and public honor, and on two occasions married rich widows. Born in colonial Woburn, Massachusetts, Thompson (1753–1814) actively aided the Loyalist Party during the American Revolution. He fled to England in 1783 and later was knighted by King George III. For many years he served the Elector of Bavaria, who granted him the title "Count Rumford of the Holy Roman Empire," even while Thompson continued to spy for the British.

In all Thompson was a keen observer of natural phenomena. While in charge of manufacturing arms for the Bavarian army he noticed that boring cannon produced large quantities of heat—heat that had to be carried away from the cannon by running water. He also determined that while the boring converted a single piece of metal stock into numerous metal flakes, the specific heat of the flakes remained identical to that of the stock. Rumford could not believe that indefinitely large amounts of a supposed ungenerate and indestructible caloric could be liberated from the metal without significantly changing its properties.

In his own words:

> It was by accident that I was led to make the experiment of which I am about to give account. . . . Being engaged lately in superintending the boring of cannons in the workshops of the military arsenal at Munich, I was struck by the considerable degree of heat that a brass gun acquires in a short time in being bored, and with the still higher temperature (much higher than that of boiling water, as I found by experiment) of the metallic chips separated from it by the borer. . . .
>
> It is hardly necessary to add, that anything which any insulated body, or system of bodies, can continue to furnish without limitation, cannot possibly be a material substance; and it appears to me to be extremely difficult, if not quite impossible, to form any distinct idea of anything, capable of being excited and communicated, in the manner the Heat was excited and communicated in these Experiments, except it be MOTION. ("Inquiry Concerning the Source of the Heat Which Is Excited by Friction," *London Philosophical Transactions*, 1798)

Evidently, the visible motion of boring a cannon produces an invisible motion in the smallest parts of the metal—a motion Rumford called heat.

More important than either Rumford's speculation on the nature of heat ("except it be MOTION") or his claim that heat can be generated without limititation is his observation that the work of boring cannon causes a "considerable degree of heat." It was already known, at least from the time of Joseph Black's somewhat earlier investigations, that heat transferred to a body increases its temperature. Now Rumford had shown that work done on a system can produce the same effect.

4.2 Joule's Experiments

Julius Robert von Mayer (1814–1878) may have been the first to infer the correct relationship between heat and work. Certainly he was the first to envision the first law of themodynamics and its great variety of applications. However, it was a series of experiments performed by the English brewer James Prescott Joule (1818–1889) that, taken as a whole, succeeded in turning Rumford's primarily qualitative observations into a quantitative law of behavior—the first law of thermodynamics. For three decades, beginning in 1842, Joule performed a series of carefully controlled and increasingly precise laboratory versions of Rumford's crude cannon-boring experiment.

In each experiment Joule measured the amount of work performed on a system through an adiabatic boundary and compared it with the amount of heat transferred to the system that caused the same temperature increase. Figure 4.1 illustrates one of his experiments. As the mass, m, falls through a distance, h, the string turns a spindle attached to paddles that stir the liquid contained within an adiabatic boundary and performs work, mgh, on the fluid.

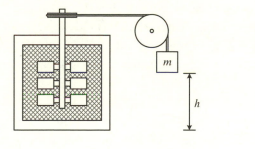

FIGURE 4.1   Joule's apparatus: a mass m falls through a distance h and performs work mgh by turning a spindle attached to a paddle wheel that is immersed in a liquid surrounded by an adiabatic boundary. Here g is the acceleration of gravity.

In some of Joule's experiments the liquid stirred was water, in some oil, and in others mercury. In some experiments work was supplied by stirring, in some by rubbing one surface on another, and in some by a current of electric charge falling through a potential difference. In each experiment the work performed could be denominated in terms of a unit of force times a unit of distance. No wonder the Systeme International (SI) unit of work is called the joule (J). In all his experiments Joule found that the ratio of work performed to the heat transferred where each causes the same increase in temperature is close to the same number. Its current value, 4.186 J/cal, is known as the *mechanical equivalent of heat*. (See Problems 4.1–4.2.)

4.3 The First Law of Thermodynamics

From Joule's experiments to the first law of thermodynamics is but a short step. Since heat, Q, transferred to and work, W, done on a system may produce the same change of state, each must be one part of a single quantity. Since work done on a purely mechanical system is known to increase its mechanical energy, it is natural to identify the sum $Q + W$ introduced into a thermodynamic system as an increase of its *internal energy*, ΔE. For this reason one common verbal expression of the first law of thermodynamics is "energy is conserved if heat is taken into account."

Algebraically the first law of thermodynamics states that

$$Q + W = \Delta E. \tag{4.1}$$

Because heat, Q, work, W, and the change in internal energy, ΔE, are, in general, signed quantities, Equation (4.1) goes beyond recapitulating the bare facts of Joule's experiments. The work may be done on or by the system or not at all ($W > 0$, $W < 0$, or $W = 0$) and by any means (mechanical, electrical, or magnetic); heat may be transferred to or rejected from the system or not at all ($Q > 0$,

$Q < 0$, or $Q = 0$); internal energy may increase, decrease, or remain the same ($\Delta E > 0$, $\Delta E < 0$, or $\Delta E = 0$); and the work, W, and heat transfer, Q, may be simultaneous or successive. Adiabatic ($Q = 0$), diathermal ($Q \neq 0$), work-allowing ($W \neq 0$), and work-prohibiting ($W = 0$) boundaries help realize these possibilities. In each process the algebraic sum of heat transferred to and work performed on the system $Q + W$ changes the internal energy by ΔE.

The discovery of the first law of thermodynamics allowed scientists to redefine the calorie as a conventional quantity denominated in energy units. Before the experiments of Joule, heat transfer was merely considered to be the transfer of something, possibly the caloric, possibly a form of motion. After Joule, heat transfer was recognized as simply one means of energy transfer. Accordingly, one calorie could be redefined in terms of units of energy—as exactly 4.186 J. However, the historically prior, water-standard definition of the calorie described in Chapter 3 remains consistent with all known thermodynamic facts. (See Problems 4.3–4.6.)

4.4 Thermodynamic Cycles

Among the many applications of the first law the most important are to cycles. A sequence of interactions that returns a system to its initial state is a thermodynamic *cycle*. These interactions may include work and heat transfer. The first law of thermodynamics [Eq. (4.1)] applied to a cycle reduces to

$$Q + W = 0 , \tag{4.2}$$

where here, as before, Q and W stand for signed quantities that are positive when heat is absorbed by and when work is performed on the system.

Numerous engineered and natural processes are composed of cycles. Machines designed to operate indefinitely do so by repeating the same cycle indefinitely. So also do biological systems that main-

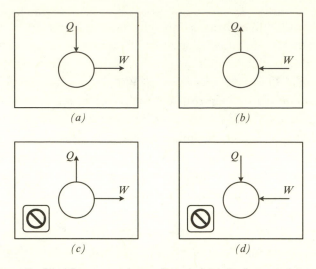

FIGURE 4.2 Cycles. The two cycles at the top—(a) the heat engine cycle and (b) the refrigerator cycle—are allowed by the first law when $Q = W$. The two cycles at the bottom (c and d) are not allowed by the first law for any values where $Q > 0$ and $W > 0$.

tain homeostasis. In principle, a cycle leaves a system unchanged while changing its environment. Whatever is accomplished in the environment in one cycle, that is, heat transferred to or from or work performed on or by, is accomplished in double amounts in two cycles and in triple amounts in three cycles.

In Figure 4.2 and throughout Chapters 5 and 6 I employ traditional diagrams to denote the operation of a cycle. Accordingly, a circle stands for a system that has gone through one or more complete cycles. Everything outside the circle is the system's environment. Arrows indicate the direction of an energy flow into or out of the system. What before, in purely algebraic formulations of the first law of thermodynamics, were signed quantities Q and W are in these diagrams mere magnitudes. Thus, Q and W denote the amount while arrows denote the direction of energy transfer. Unsigned quantities Q and W always label diagrams; in other circumstances, for generality, Q and W are signed. Thus, the context is important.

Figure 4.2 illustrates, in this diagrammatic language, two possible cycles: (a) a heat engine cycle that extracts net heat from its environment and produces work and (b) a refrigerator cycle that consumes work and rejects net heat into its environment; in addition, it illustrates two impossible cycles, (c) and (d). In each possible cycle the system absorbs or rejects as much heat as work produced or consumed so that $Q = W$. Each impossible cycle, marked with a $\oslash$ symbol, necessarily violates the first law of thermodynamics for any quantities Q and W where $Q > 0$ and $W > 0$.

Cycles in which net heat input or output does not balance the work produced or consumed can be reduced to a few types whose diagrams are easily recognized as violating the first law. Suppose, for instance, a heat engine cycle of the form shown in Figure 4.2a produces more work, W, than heat consumed, Q, so that $W > Q$. We could combine this prohibited heat engine cycle with an allowed refrigerator cycle of form illustrated in Figure 4.2b that consumes work, W', and rejects the same amount of heat, so that $Q' = W'$. Suppose we adjust both cycles (as described in the next section) until the heat absorbed from the environment by the heat engine equals the heat rejected to the environment by the refrigerator, that is, so that $Q' = Q$. Then, the combined system composed of the heat engine and the refrigerator peforms net positive work, $W - W' > 0$, on the environment without exchanging net heat with the environment. However convenient, this and other *one-flow cyclic processes*, diagrammed in Figure 4.3, necessarily violate the first law of thermodynamics.

Of course, one-flow processes that only produce work without absorbing or rejecting heat do exist: examples are a discharging capacitor or a piston released and pushing against its environment. And there are one-flow processes that simply release net heat, for instance, a cup of tea cooling down to room temperature. But none of these return the systems (capacitor, piston, tea) to their initial states. None of these processes are cycles.

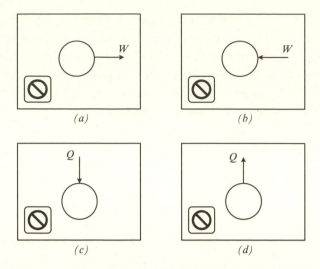

FIGURE 4.3 Four one-flow cycles prohibited by the first law of thermo-dynamics. Here $Q > 0$ and $W > 0$.

4.5 Cycle Adjustment

An *adjustment* that makes a single quantity in one cycle equal to a quantity of the same kind in another cycle can always be made. Consider, as before, a heat engine cycle that absorbs heat Q where $Q > 0$ and produces work W where $W > 0$ and a refrigerator cycle that rejects heat Q' where $Q' > 0$ and consumes work W' where $W' > 0$. By repeating the heat engine cycle n times and the refrigerator cycle n' times, where the integers n and n' are chosen in such a way that their ratio, n/n', is as close as desired to the ratio Q'/Q, we can compose a third cyclic device out of the combined heat engine and refrigerator that exchanges no net heat with the environment, since $nQ - n'Q' = 0$, yet formally exchanges net work, $n'W' - nW$, with the environment. In similar fashion a single adjustment can always equalize any two, but no more than two, quantities of the same type, Q or W, in different cycles. Cycle adjustment prepares us for a similar if more complex analysis of the second law of thermodynamics, a task taken up in Chapter 5. (See Problems 4.7–4.8.)

Chapter 4 Problems

4.1 *Mechanical Equivalent of Heat*. Rumford observed that when one horse turned a cannon bore for about 2.5 hr the temperature of 27 lb of water that was initially ice cold (i.e., near 0°C) rose to near boiling hot (i.e., near 100°C). Given that one horsepower is 33,000 ft-lb/min, that 1 kg weighs 2.2 lb, and that 1 in. = 2.54 cm, what is the mechanical equivalent of heat in joules per calorie implied by these rough numbers?

4.2 *Waterfall*. William Thomson (later known as Lord Kelvin) told the story (possibly apocryphal) of a holiday in Switzerland in which he saw a young man approaching him with what looked like a walking stick gingerly balanced in vertical position. It was James Joule on his honeymoon! He had brought along a huge thermometer in order to compare the water temperature at the top and at the bottom of an 800-ft waterfall he and his bride were visiting. How much hotter at the bottom than at the top should Joule have expected the water to be?

4.3 *Energy Balance*. A system absorbs 45 cal of heat and performs 200 J of work. What is the net change in internal energy, ΔE, of the system? Recall that ΔE may be positive or negative and that, in principle, $\Delta E = E_f - E_i$, where E_f and E_i are, respectively, the amounts of internal energy after the process and before the process.

4.4 *Stirring*. A liquid is stirred in a well-insulated container until its temperature rises.
(a) Has heat been transferred to the liquid?
(b) Has work been performed on the liquid?
(c) What is the sign of ΔE?

4.5 *Free Expansion*. A well-insulated container with rigid walls is divided into two parts separated by a partition that can be easily ruptured. A gas is initially contained in only one of these parts; the other part is completely evacuated. The partition is ruptured and the gas fills both parts

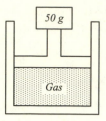

FIGURE 4.4 Cylinder and piston head containing a gas. (See Problem 4.6.)

of the system. Does the internal energy of the system—that is, the gas—increase, decrease, or remain the same?

4.6 *Friction/Dissipation.* A volume of gas is contained in a well-insulated cylinder with a well-insulated piston head as depicted in Figure 4.4. The massless piston head may move but only by overcoming 0.2 newtons of kinetic friction. A 50-g mass is placed on top of the piston head. The piston head moves outward a distance of 25 cm. Ignore the pressure exerted by the atmosphere.

(a) What is the amount of work performed by the gas during this expansion?

(b) If we consider the gas, the cylinder, and the piston head to be part of the system, what is the change in internal energy, ΔE, of the system?

4.7 *First Law Equivalent.* The first law of thermodynamics can be formulated as the following statement of impossibility: "It is impossible to devise a cycle that has no effect other than the performance of work on or by the environment."

(a) Use an indirect proof to show that this impossibility version of the first law, the existence of heat engine and refrigerator cycles diagrammed in Figure 4.2a–b, and the possibility of adjusting and combining cycles, as explained in Section 4.5, together lead to the denial of the one-flow heat cycles diagrammed in Figure 4.3c–d. (S)

(b) Show that the impossibility of the one-flow heat cycles diagrammed in Figure 4.3c–d, the existence of the heat engine and refrigerator cycles diagrammed in Figure 4.2a–b, and the possibility of adjusting and combining cycles, as explained in Section 4.5, lead to this impossibility statement of the first law.

Parts (a) and (b) combined prove that, given the possibility of adjusting and combining cycles and the existence of heat engine and refrigerator cycles, the impossibility version of the first and the denial of the one-flow heat cycles diagrammed in Figure 4.3c–d are logically equivalent versions of the first law of thermodynamics.

4.8 *Cycle.* A system undergoes the following four-stage cyclic process: In stage (1) the system absorbs 226 cal of heat and does 50 J of work; in (2) the system adiabatically does 30 J of work; and in (3) the system rejects 100 cal of heat while the environment does 80 J of work on the system. Stage (4) is also adiabatic. Is work done on or by the system in stage (4)? How many joules of work?

The Second Law

....

5.1 Sadi Carnot
....

Sadi Carnot (1796–1832) was the brilliant son of an illustrious father. Lazare Carnot had appointed Napoleon to his first independent command and organized the fourteen armies that conquered Europe. As an 18-year-old military student, Sadi participated in the defense of Paris. But by 1824 Napoleon was dead, having been imprisoned on a small island in the mid-Atlantic; Lazare too had died in exile; and Sadi was living a quiet life in Paris studying physics and economics. That year Sadi Carnot laid the foundations for an idea more consequential than the Napoleonic conquest of Europe—an idea we now call the second law of thermodynamics.

Carnot's immediate goal in his essay "Reflections on the Motive Power of Heat and on Machines Fitted to Develop That Power" was to identify a theoretical limit on the efficiency of heat engines. He found the origin of that limit in his observation that "the production of motion in steam engines is always accompanied by a circumstance on which we should fix our attention. This circumstance is the re-establishing of equilibrium in the caloric; that is, its passage from a body in which the temperature is more or less elevated to another in which it is lower." Carnot was so impressed with this idea—that all heat engines receive heat from a hot body and reject waste heat to a cold body—that he repeated it seven times in seven consecutive paragraphs of the "Reflections."

Carnot worked without benefit of the first law of thermodynamics. After all, his "Reflections" appeared some 20 years before Joule's convincing experiments of the 1840s. Instead of the first law, Carnot favored, as did many others in 1824, the theory of caloric, according to which heat is a substance that can neither be created nor destroyed.

Not until 1850 did Rudolph Clausius harmonize Carnot's idea with the conservation of energy and call the two, respectively, the second and first laws of thermodynamics. Interestingly, Clausius's ordering of the two laws of classical thermodynamics reverses the order of their discovery. In this chapter we explore, as did Sadi Carnot in 1824, the consequences of the second law without assuming the first law. In this way we isolate the logical content of the second law and recreate the intellectual context of its discovery.

5.2 Statements of the Second Law

Figure 5.1 illustrates the essential feature of Carnot's idea: the simplest possible heat engine operating in a cycle extracts heat, Q_H, from a hot heat reservoir, rejects waste heat, Q_C, to a colder one, and produces work, W. Here and elsewhere I use the term *heat reservoirs* instead of Carnot's hot and cold bodies. By definition, heat can be added to or extracted from a heat reservoir without changing its temperature. A heat reservoir has infinite heat capacity.

Figure 5.2 illustrates heat engines that, according to Carnot, are too simple to be possible. As before, forbidden cyclic processes are identified with a $\oslash$ symbol. Of course, Carnot would have found the explicit prohibition of engines 5.2b and 5.2c unnecessary, even

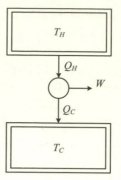

FIGURE 5.1   Carnot's simplest heat engine, operating in a cycle, extracts heat, Q_H, from a heat reservoir at temperature T_H; performs work, W; and rejects waste heat, Q_C, to a reservoir at a lower temperature, where $T_C < T_H$. Carnot, the caloricist, would assume that $Q_H = Q_C$.

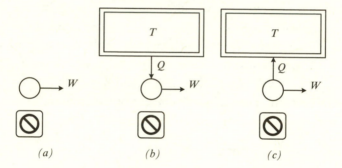

FIGURE 5.2   Heat engines too simple to be possible. Carnot's second law forbids each because each cycle produces work by exchanging heat with fewer than two heat reservoirs.

apart from his new principle, since each violates the conservation of caloric. Similarly, today we find the impossibility of engines 5.2a and 5.2c obvious because each of these necessarily violates the first law of thermodynamics. However, if we assume neither the conservation of energy nor the conservation of caloric—as we do in this chapter—all three statements of impossibility are necessary to express Carnot's principle. Cast into negative form that principle becomes a version of the second law of thermodynamics:

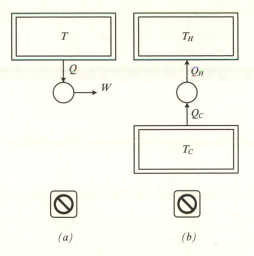

FIGURE 5.3 (a) Heat engine forbidden by Thomson's second law. (b) Heat flow process forbidden by Clausius's second law. Here, we intentionally strip Thomson's and Clausius's statements of any first-law content—that is, we assume neither that $Q = W$ in (a), nor that $Q_H = Q_C$ in (b).

> *A process whose only result is to exchange heat with fewer than two different heat reservoirs and produce work is impossible.*

Carnot himself never articulated this version of the second law, but this statement so closely follows from Carnot's own words that I take the liberty of calling it *Carnot's second law.*

Other, better known statements of the second law originated with William Thomson (1824–1907) and Rudolph Clausius (1822–1888). According to *Thomson's second law,*

> *A process whose only result is to extract heat from one heat reservoir and produce work is impossible,*

and, according to *Clausius's second law,*

> *A process whose only result is to extract heat from one heat reservoir and reject heat to another hotter heat reservoir is impossible.*

The Thomson and Clausius statements of the second law are diagrammed in Figure 5.3.

Each of these three statements of the second law—Carnot's, Thomson's, and Clausius's—prohibits certain cyclic processes. Each statement is verbally distinct from the other two. It is clear, however, that Thomson's second law, because it forbids only a subset of those forbidden by Carnot's second law, restricts the world less severely than Carnot's. And I will show that Carnot's second law is logically equivalent to Clausius's second law. (See Problem 5.1.)

5.3 Equivalence and Inequivalence

In order to demonstrate the equivalence and inequivalence of different versions of the second law, we explicitly assume that certain heat engines and heat flow processes are possible. These include the simplest heat engine of Carnot, a simple refrigerator cycle, and heat flow from a hotter heat reservoir to a colder one. In one form or another each of these cyclic processes has often been approximated in technology or observed in nature. All three are illustrated in Figure 5.4.

We can now show that Carnot's and Clausius's versions of the

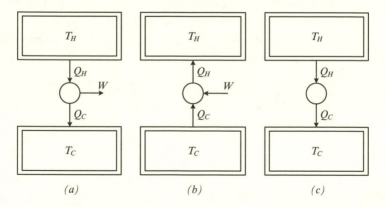

FIGURE 5.4   Explicitly allowed cyclic processes: (a) simple heat engine; (b) simple refrigerator; (c) heat flow from a hotter to a colder heat reservoir.

second law are logically equivalent. We do this by establishing that Clausius's second law implies Carnot's second law (Claim 1) and that Carnot's second law implies Clausius's second law (Claim 2). If we can establish both Claim 1 and Claim 2, we establish the logical equivalence of the Carnot and Clausius versions of the second law.

Claim 1 is that Clausius's second law implies Carnot's second law. We prove Claim 1 by proving its contrapositive, namely, that to deny Carnot's second law leads to a denial of Clausius's second law. Each of the three different ways of violating Carnot's second law leads to a violation of Clausius's second law. Suppose, for instance, we construct, as diagrammed in Figures 5.2a and 5.5a, cyclic heat engine 1 that produces work W without exchanging heat with its environment. We are free to adjust the work W produced by heat engine 1 so that it supplies the work consumed W' by allowed refrigerator 2 in Figure 5.5a. The net result of the combined cyclic operation of the supposed heat engine and the explicitly allowed refrigerator, cycle 1&2, is to extract heat Q_C from a reservoir at temperature T_c and reject heat Q_H to a reservoir at higher temperature T_H without consuming work—a clear violation of Clausius's second law.

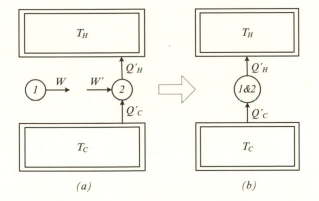

FIGURE 5.5 (a) A supposed heat engine that violates Carnot's second law and an allowed refrigerator. (b) The combined heat transfer resulting from the adjustment $W = W'$. Heat engine 1&2 violates Clausius's second law.

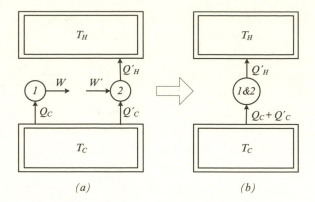

FIGURE 5.6 (a) A supposed heat engine, 1, that violates Carnot's second law and an allowed refrigerator, 2. (b) The combined heat transfer resulting from the adjustment $W = W'$. Heat engine 1&2 violates Clausius's second law.

The other ways of violating Carnot's second law—that is, those that suppose the existence of the engines diagrammed in Figure 5.2b and c—similarly lead to violations of Clausius's second law. One has only to closely follow the pattern of proof already established—each time adjusting the work W produced by the supposed heat engine to provide the work W' needed to run an allowed refrigerator. In each case the combined engine can be made to violate Clausius's second law. The logic of these deductions is made clear in Figures 5.6 and 5.7. Consequently, the contrapositive of Claim 1—that a violation of Carnot's second law leads to a violation of Clausius's second law—is established. Therefore, Claim 1 is established.

Next we prove Claim 2—that Carnot's second law leads to Clausius's second law. We again do so by proving its contrapositive—that violating Clausius's second law leads to a violation of Carnot's second law. If we can violate Clausius's second law, then we can arrange for heat Q_C to be extracted from a reservoir at temperature T_c and heat Q_H to be rejected to a hotter reservoir at temperature T_H without consuming work. This supposed cyclic heat flow, diagrammed in Figures 5.3b and 5.8a, can be adjusted to ex-

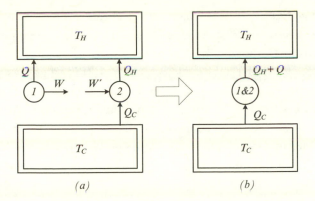

FIGURE 5.7 (a) A supposed heat engine, 1, that violates Carnot's second law and an allowed refrigerator, 2. (b) The combined heat transfer resulting from the adjustment $W = W'$. Heat engine 1&2 violates Clausius's second law.

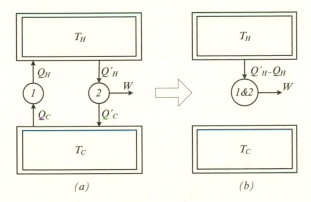

FIGURE 5.8 (a) A supposed heat flow process, 1, that violates Clausius's second law and an allowed heat engine, 2. (b) The combined heat engine resulting from the adjustment $Q_C = Q'_C$. Heat engine 1&2 violates Carnot's second law.

tract exactly that heat deposited in the T_c reservoir Q'_C by the explicitly allowed heat engine diagrammed in Figure 5.4a and in Figure 5.8a. The net result of combining these cycles is to exchange heat with a single reservoir at temperature T_H and produce work W as shown in Figure 5.8b. Of course, we do not know whether the heat exchanged $Q'_H - Q_H$ with the reservoir at temperature T_H in Figure 5.8b is positive, negative, or zero, but each of these three possibilities violates Carnot's second law.

Since we have established both Claim 1, that Clausius's second law implies Carnot's second law, and Claim 2, that Carnot's second law implies Clausius's second law, we have established the logical equivalence of these two versions of the second law. We leave it to the reader to show that a similar attempt to prove the equivalence of the Thomson and Clausius formulations fails, as it must. (See Problem 5.2.)

5.4 Reversible Heat Engines

In his quest to conceptualize heat engine operation Carnot invented the idea of a *reversible cycle*. Reversible cycles and processes are those that proceed indefinitely slowly (that is, quasistatically) in order that the system state always remain in equilibrium and also without friction or internal dissipation. An important feature of reversible cycles and processes is that they are uniquely calculable. But here we exploit only their eponymous—that is, naming—character: a reversible process can proceed in reverse.

Reversing a reversible heat engine transforms it into a reversible refrigerator: the work produced by the engine is now consumed by the refrigerator, the heat extracted by the engine from the hotter reservoir is now rejected to it, and the heat rejected by the engine to the colder reservoir is now extracted from it. A reversible cycle that is run once in one direction and then run once again in the other direction would completely undo all that it has accomplished. Of course, reversible heat engine and refrigerator cycles are idealizations; they can be approximated but never fully realized in practice. Furthermore, as we shall see, some simple cycles are inherently irreversible.

The following theorems on reversible heat engines that operate between two temperature reservoirs, also called *Carnot cycles*, are consequences of their eponymous property and of either the Carnot or the Clausius version of the second law but do not require the first law of thermodynamics. Sadi Carnot proved Proposition I (sometimes called *Carnot's theorem*) and assumed the truth of Propositions II and III.

Carnot was primarily interested in the greatest efficiency of the simplest cyclic heat engine. This *efficiency* is the ratio of the work the engine produces, W, to the heat it absorbs from the hotter of its two reservoirs, Q_H, that is W/Q_H. Recall that, apart from the first law, which Carnot did not accept, this efficiency has irreducible units of calories/joule. Carnot asked, What is the greatest efficiency of such a reversible heat engine and on what does this efficiency depend or not depend?

PROPOSITION I (Carnot's theorem): *No heat engine operating between two heat reservoirs can be more efficient than a reversible heat engine.*

STYLE OF PROOF: By contradiction

PROOF:

Suppose the contradiction of Proposition I—that is, an irreversible heat engine *Ir* is more efficient than a reversible one *R* operating between the same two heat reservoirs. See Figure 5.9a.

According to this supposition and the definition of heat engine efficiency $W^R/Q_H^R < W^{Ir}/Q_H^{Ir}$.

Adjust the two engines so that $Q_H^R = Q_H^{Ir}$.

Then $W^R < W^{Ir}$.

Reverse the reversible engine *R* as shown in Figure 5.9b and combine the two engines into one as shown in Figure 5.9c.

The combined engine produces work $W^{Ir} - W^R > 0$ and exchanges heat $Q_C^{Ir} - Q_C^R$ with one heat reservoir at temperature T_C.

Since this deduction contradicts Carnot's second law, one of the previous assumptions must be false.

Since all assumptions except the original one are unimpeachable, the contradiction of Proposition I must be false.

Therefore, Proposition I has been proven.

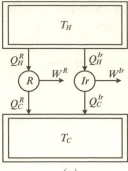

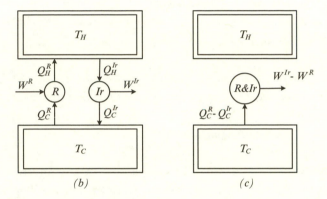

FIGURE 5.9 Illustrations for the proof of Proposition I (Carnot's theorem). (a) A reversible engine R and an irreversible heat engine Ir operating between the same two heat reservoirs. (b) The reversible engine R is reversed. (c) The adjustment $Q_H^R = Q_H^{Ir}$ is made and the two engines combined.

PROPOSITION II: *All reversible heat engines operating between the same two heat reservoirs have the same efficiency.*

The proof of Proposition II follows the pattern established by the proof of Proposition I and is left as an exercise. (See Problem 5.3.) An immediate consequence of Proposition II is that the efficiency of a reversible heat engine operating between two heat reservoirs is a function $\varepsilon(T_H, T_C)$ of the reservoir temperatures T_C and

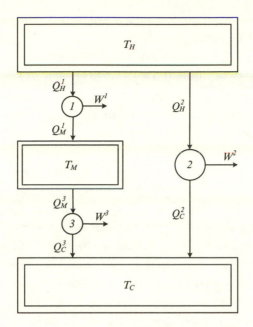

FIGURE 5.10 Reversible heat engines used in the proof of Proposition III.

$T_H > T_C$ alone. Carnot and others recognized the importance of the function $\varepsilon(T_H, T_C)$. They sought, and sometimes assumed, its form but were never able to deduce $\varepsilon(T_H, T_C)$ from the second law alone.

PROPOSITION III: *The efficiency of a reversible heat engine operating between two heat reservoirs decreases as the lesser of the two temperatures increases.*

STYLE OF PROOF: Direct

PROOF:

Consider two reversible heat engines, 1 and 2, operating between a common reservoir with temperature T_H and different colder reservoirs with temperatures T_M and T_C where $T_H > T_M > T_C$, as shown in Figure 5.10.

Construct a third reversible heat engine 3 operating between the reservoirs with temperatures T_M and T_C.

Adjust engines 1 and 3 so that $Q_M^1 = Q_M^3$. This adjustment combines engines 1 and 3 into a single engine 1&3 from which the T_M reservoir is effectively eliminated.

Adjust engines 1&3 and 2 so that $Q_H^1 = Q_H^2 (\equiv Q_H)$.

According to Proposition II, $(W^1 + W^3)/Q_H = W^2/Q_H$.

Therefore, $W^2/Q_H \geq W^1/Q_H$, that is, $\varepsilon(T_H, T_C) \geq \varepsilon(T_H, T_M)$ the equality being realized only when $T_C = T_M$.

Therefore, Proposition III has been proven.

William Thomson realized, in 1848, that Proposition III allowed one to use the efficiency of a reversible heat engine operating between two heat reservoirs as a thermometric property. A temperature scale based upon this thermometric property would be independent of the working fluid or the design of any particular reversible heat engine and in this sense would be *universal* or *absolute*. Although interesting, absolute temperature scales based on the second law alone are not as convenient as ones that depend upon both the first and second laws of thermodynamics. (See Problems 5.3–5.4 and Appendix B.)

5.5 Refrigerators and Heat Pumps

Refrigerators and heat pumps are heat engines with special purposes. A refrigerator transfers heat from a cold body to a hot body in order to maintain the temperature of the cold body, while a heat pump transfers heat from a cold body to a hot body in order to maintain the temperature of the hot body.

Let's consider a refrigerator operating between two reservoirs at temperatures T_H and $T_C < T_H$. The best refrigerator will extract the most heat Q_C from the colder reservoir for a given cost, that is, for a given work, W, necessary to run the refrigerator. For this reason engineers have invented a refrigerator *coefficient of performance*, Q_C/W, that quantifies refrigerator operation; the better the refrigerator performs the higher Q_C/W. Interestingly, no refrigerator operating between two heat reservoirs can have a higher coefficient of performance than one run by a reversible engine, that is, a reversed Carnot engine. The proof follows a pattern similar to that followed in proving Carnot's theorem—that no heat engine operating between two heat reservoirs can have a higher efficiency than a reversible engine—and is outlined in Problem 5.5. (See Problems 5.5–5.6.)

Chapter 5 Problems

....

5.1 *Faulty Second Law.* Carefully consider the following formulation of the second law: *A process that extracts heat from one heat reservoir and produces work is impossible.*
(a) How is this different from Thomson's second law?
(b) Describe a counter-example to this faulty formulation of the second law.
[Hint: Review Sections 4.4 and 5.2.]

5.2 *Clausius's Second Law.* Prove that the following processes lead to a violation of Clausius's second law:
(a) A cyclic process that absorbs heat from a reservoir and produces work as shown in Figure 5.2b (S).
(b) A cyclic process that rejects heat to a reservoir and produces work as shown in Figure 5.2c.
[Hint: See Figures 5.6 and 5.7.]

5.3 *Proposition II.* Prove Proposition II using the same methods used in proving Proposition I.

5.4 *Proposition III.* Prove Proposition III without appealing to Proposition I (Carnot's Theorem) or Proposition II.
[Hint: Construct an indirect proof based on contradicting Carnot's second law.]

5.5 *Ideal Refrigerator Performance.* Prove that no refrigerator operating between two heat reservoirs has a higher coefficient of performance Q_C/W than one run by a reversible engine. To do so, show that the negation of this statement—that the coefficient of performance Q'_C/W' of an irreversible refrigerator is higher than the coefficient of performance Q_C/W of a reversible refrigerator operating between the same two heat reservoirs—leads to a violation of the second law.

[Hint: Diagrams may be helpful. By supposition $Q'_C/W' > Q_C/W$. First reverse the reversible refrigerator, then adjust $Q'_C = Q_C$ so that the supposition leads to $W > W'$. This violates Carnot's second law.]

5.6 *Ideal Heat Pump Performance.* The coefficient of performance of a heat pump is Q_H/W. Prove that no heat pump operating between two heat reservoirs has a higher coefficient of performance than one run by a reversible engine. Follow the pattern of proof established in Problem 5.5.

SIX

The First and Second Laws

6.1 Rudolph Clausius

In 1850 the young Rudolph Clausius (1822–1888) solved a problem
worthy of his vocation in theoretical physics: harmonizing Carnot's
analysis of heat engines and Joule's discovery of energy conservation. Clausius first remarked on the difficulty of either retaining all
or abandoning all of Carnot's analysis. Then he discerned a middle
course.

> A careful examination shows that the new method does not
> stand in contradiction to the essential principle of Carnot, but
> only to the subsidiary statement that [in heat engine operation]
> *no heat is lost*, since in the production of work it may very well
> be the case that at the same time a certain quantity of heat is
> consumed and another quantity transferred from a hotter to a
> colder body, and both quantities of heat stand in definite rela-
> tion to the work that is done.

In abstracting Carnot's "essential principle" (that is, the second law) from its inessential expression and replacing the conservation of caloric with Joule's "new method" (that is, with the first law), Clausius created classical thermodynamics. The two laws of classical thermodynamics, in combination, imply the existence of an absolute or thermodynamic temperature scale, the so-called Kelvin scale, and validate a theorem first proved by Clausius and appropriately called *Clausius's theorem*. These concepts transform the mainly verbal and logical expressions of the first and second laws we have seen thus far into a powerful mathematical theory. We explore this transition from words to mathematics in this chapter and the next.

6.2 Thermodynamic Temperature

The efficiency, ε, of a heat engine operating between two heat reservoirs is defined by

$$\varepsilon = \frac{W}{Q_H}, \tag{6.1}$$

where W is the work produced by the engine and Q_H is the heat absorbed by the engine from the hotter reservoir. According to the first law, the heat absorbed yet not transformed into work,

$$Q_C = Q_H - W, \tag{6.2}$$

is rejected to the colder reservoir. Thus, the ratio of heat rejected to heat absorbed, Q_C/Q_H, is related to the efficiency of the engine, ε, by

$$\frac{Q_C}{Q_H} = 1 - \varepsilon. \tag{6.3}$$

According to Carnot's theorem all reversible heat engines operating between reservoirs with the same two temperatures, T_C and

T_H—that is, all Carnot cycles—are described by the same efficiency function, $\varepsilon(T_C,T_H)$. Consequently, the heat ratio function

$$\frac{Q_C}{Q_H} = f(T_C,T_H), \tag{6.4}$$

where $f(T_C,T_H) \equiv 1 - \varepsilon(T_C,T_H)$, also depends only upon the two temperatures, T_C and T_H, and not otherwise upon the design or composition of the reversible heat engine.

This heat ratio function $f(T_C,T_H)$ has certain easily discovered properties. For instance, since according to Equations (6.1) and (6.2) and the convention that $Q_H > 0$, $Q_C > 0$, and $W > 0$,

$$0 \le f(T_C,T_H) \le 1. \tag{6.5}$$

We also find, from the following reasoning, that $f(T_C,T_H)$ obeys the cyclic property

$$f(T_1,T_2)f(T_2,T_3) = f(T_1,T_3) \tag{6.6}$$

and, consequently, that

$$f(T,T) = 1. \tag{6.7}$$

To derive Equation (6.6) consider three heat reservoirs with ascending temperatures, $T_1 \le T_2 \le T_3$, and two Carnot cycles operating between the reservoirs. The heat ratio of the Carnot cycle operating between temperatures T_1 and T_2 is

$$\frac{Q_C^1}{Q_H^2} = f(T_1,T_2), \tag{6.8}$$

and the heat ratio of a Carnot cycle operating between T_2 and T_3 is

$$\frac{Q_C^2}{Q_H^3} = f(T_2,T_3), \tag{6.9}$$

where the numerical superscripts denote the reservoir from which heat is absorbed or to which heat is rejected. By adjusting the heat extracted, Q_H^2, from the middle (T_2) reservoir by the first Carnot cycle and the heat rejected, Q_C^2, into the middle (T_2) reservoir by the second Carnot cycle so that $Q_C^2 = Q_H^2$, we combine these two Carnot cycles into one whose only consequence is to extract heat Q_H^3 from the hottest (T_3) reservoir and reject heat Q_C^1 to the coldest (T_1) reservoir. Thus, the combined cycle is also a Carnot cycle with its own heat ratio

$$\frac{Q_C^1}{Q_H^3} = f(T_1, T_3). \tag{6.10}$$

Since $Q_H^2 = Q_C^2$, Equations (6.8)–(6.10) imply

$$\frac{Q_C^1}{Q_H^2} \cdot \frac{Q_C^2}{Q_H^3} = \frac{Q_C^1}{Q_H^3}, \tag{6.11}$$

which is equivalent to the cyclic property $f(T_1, T_2)f(T_2, T_3) = f(T_1, T_3)$, that is, to

$$f(T_1, T_2) = \frac{f(T_1, T_3)}{f(T_2, T_3)}. \tag{6.12}$$

Note that whatever the functional form of $f(T_1, T_3)$ or, equivalently, $f(T_2, T_3)$, it allows the dependence on T_3 to disappear from the right-hand side of Equation (6.12).

The only function $f(T_C, T_H)$ observing the properties stated in Equations (6.5)–(6.7) has the structure

$$f(T_C, T_H) = \frac{\phi(T_C)}{\phi(T_H)}. \tag{6.13}$$

Whatever the form of $\phi(T)$, Equation (6.13) allows us to adopt a *thermodynamic temperature scale* that is independent of any one system. Essentially, we define a *thermodynamic temperature* $\tau = \phi(T)$ so that

$$\frac{\tau_C}{\tau_H} = \frac{\phi(T_C)}{\phi(T_H)}. \tag{6.14}$$

Because $Q_C/Q_H = f(T_C, T_H) = \phi(T_C)/\phi(T_H) = \tau_C/\tau_H$, adopting a thermodynamic temperature scale is equivalent to adopting the heat ratio of a Carnot cycle as the thermometric property. And because $0 \leq Q_C \leq Q_H$, thermodynamic temperatures are such that $0 \leq \tau_C/\tau_H \leq 1$. The equality $\tau_C/\tau_H = 1$ obtains only when $T_C = T_H$. Furthermore, $\tau_C \to 0$ for arbitrary τ_H only as $Q_C \to 0$. Thus, thermodynamic temperatures are single-sign—that is, all are non-negative (the common choice) or non-positive and all have a common limiting value of zero. This definition of thermodynamic temperature depends upon both the first and second laws. After all, if caloric were conserved, then $Q_C = Q_H$ and the heat ratio Q_C/Q_H could not serve as a thermometric property.

Equation (6.14) allows many different thermodynamic temperature scales, each with a different-sized degree. The Kelvin degree is defined so that the temperature of the triple point of water is 273.16 K. On the other hand, the Rankine degree is, by definition, exactly 5/9 of one Kelvin degree—that is, exactly the size of a Fahrenheit degree.

Expressed in terms of thermodynamic temperatures the efficiency of a reversible heat engine operating between two heat reservoirs—that is, the efficiency of a Carnot cycle—is

$$\varepsilon(\tau_C, \tau_H) = 1 - \frac{\tau_C}{\tau_H}. \tag{6.15}$$

This efficiency obeys those theorems proposed by Carnot in the simplest possible way: it agrees with Proposition II, that all Carnot cycles have the same efficiency, and with Proposition III, that the efficiency of all Carnot cycles decreases as the colder of the two temperatures increases.

Thermodynamic temperatures similarly reduce equations of state to relatively simple forms. Because we use thermodynamic temperatures (usually the Kelvin scale) almost exclusively in what follows, we will abandon the notational distinction between thermodynamic τ and empirical temperatures T that has been adopted thus far. Henceforth, all temperatures are denoted with the common symbol T. The context indicates the kind of temperature intended—empirical or thermodynamic. (See Problems 6.1–6.6.)

6.3 Clausius's Theorem

According to Carnot's theorem (Proposition I), the efficiency, $1 - Q_C/Q_H$, of an arbitrary cyclic heat engine operating between hotter and colder heat reservoirs is less than or equal to the efficiency, $1 - T_C/T_H$, of a reversible engine operating between the same two heat reservoirs—that is,

$$1 - \frac{Q_C}{Q_H} \leq 1 - \frac{T_C}{T_H} \tag{6.16}$$

or, alternatively,

$$\frac{Q_H}{T_H} + \frac{-Q_C}{T_C} \leq 0, \tag{6.17}$$

where here, as in Chapter 5, $Q_H > 0$ and $Q_C > 0$. The equality in Equations (6.16) and (6.17) obtains only when the arbitrary cycle is reversible. Inequality (6.17) is one instance of *Clausius's theorem* or *Clausius's inequality*.

In order to represent Clausius's theorem in its most general form we return to the general custom of using the symbols Q and W to represent signed quantities associated with a particular system. Thus $Q > 0$ when the system absorbs heat, $Q < 0$ when the system gives up heat, $W > 0$ when the system produces work, and $W < 0$ when the system consumes work. In the language of this convention, Clausius's theorem states that

$$\frac{Q_H}{T_H} + \frac{Q_C}{T_C} \leq 0 \tag{6.18}$$

when the cycle operates between two heat reservoirs.

Suppose a system experiences a cyclic, but otherwise arbitrary, process that produces or consumes work, W, while exchanging heat, $Q_1, Q_2, ..., Q_i, ...Q_n$, with n reservoirs each having respective temperatures $T_1, T_2, ...T_i, ...T_n$. Again, when $W > 0$ the system produces work, when $W < 0$ the system consumes work, when $Q_i > 0$ the system absorbs heat from the ith reservoir, and when $Q_i < 0$ the system rejects heat to the ith reservoir. Clausius's theorem applied to this complex arbitrary cycle requires that

$$\sum_{i=1}^{n} \frac{Q_i}{T_i} \leq 0, \tag{6.19}$$

where the equality sign obtains only when the complex cycle is reversible. These relationships are represented in Figure 6.1, in which the arrows represent the direction of positive work and heat flow rather than the actual direction of work and heat flow in a particular case.

To prove Clausius's theorem [Eq. (6.19)] we combine the complex arbitrary cycle with n Carnot cycles operating between the n reservoirs with temperatures $T_1, T_2, ...T_i, ...T_n$ and another reservoir of arbitrary temperature T_o. Each of the n Carnot cycles exchanges heat Q_i' with each T_i reservoir, produces (or consumes) work W_i', and exchanges heat Q_{oi}' with the T_o reservoir, so that

$$\frac{Q_i'}{T_i} + \frac{Q_{oi}'}{T_o} = 0 \tag{6.20}$$

for each i. Furthermore, each Carnot cycle is adjusted so that it supplies to or extracts from each of the n reservoirs that heat extracted from or supplied to it by the original complex cycle. In this way

$$Q_i' = -Q_i \tag{6.21}$$

and the n reservoirs with temperatures $T_1, T_2, ...T_i, ...T_n$ merely transmit heat.

The result of combining the original arbitrary complex cycle and the n Carnot cycles is to create a simple cycle that extracts net

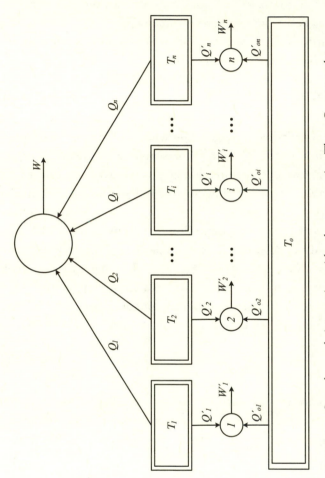

FIGURE 6.1 Complex cycle interacting with n heat reservoirs. The n Carnot cycles absorb or reject the same heat, $Q'_i = -Q_i$, rejected to or absorbed from the n heat reservoirs.

heat $\Sigma_1^n Q'_{oi}$ from the T_o reservoir and produces work $W + \Sigma_i^n W'_i$. But the Thomson and Carnot statements of the second law prohibit a cycle from absorbing heat from a single heat reservoir and producing work. Therefore, $W + \Sigma_i^n W_i \leq 0$, which, given the first law, implies

$$W + \sum_i^n W_i = \sum_{i=1}^n Q'_{oi} \leq 0. \tag{6.22}$$

Since, from Equation (6.20), each quantity $Q'_{oi} = -Q'_i T_o / T_i$ and, from (6.21), $Q'_i = -Q_i$, (6.22) becomes

$$\sum_{i=1}^n \frac{Q_i}{T_i} \leq 0, \tag{6.23}$$

where the common factor T_o has been factored from each term of the sum.

However, Equation (6.23) establishes only part of Clausius's theorem. The balance of the proof must demonstrate that the equality sign in (6.23) obtains when the original, arbitrary, complex cyclic process is reversible. If the original cyclic process is reversible, every part of the combined cycle is reversible. Reversing the combined cycle reverses the signs of all heat exchanges in (6.23) and so produces

$$\sum_{i=1}^n \frac{Q_i}{T_i} \geq 0, \tag{6.24}$$

which obtains, along with (6.23), for complex reversible cyclic transformations if and only if

$$\sum_{i=1}^n \frac{Q_i}{T_i} = 0. \tag{6.25}$$

This completes the proof of Clausius's theorem.

Because Clausius's theorem [Eq. (6.23)] does not restrict the number n, each finite heat exchange Q_i with a heat reservoir of temperature T_i can be subdivided into an indefinitely large number of infinitesimal interactions dQ. Thus, Clausius's theorem can also be expressed as

$$\oint \frac{dQ}{T} \leq 0, \qquad (6.26)$$

where the circle on the integral sign reminds us that the infinitesimal heat quantities dQ are those absorbed by the system in the course of a complete cycle. Recall also that the temperature T in (6.26) is that of the reservoir with which the system interacts and not necessarily the temperature of the system. The equality sign is realized only when the cycle is reversible. In that case T is the common temperature of the system and the reservoirs with which it exchanges heat. Finally, the work W produced by the system experiencing the cyclic transformation is, according to the first law, equal to the total heat absorbed by the system so that, in this notation,

$$W = \oint dQ. \qquad (6.27)$$

(See Problems 6.7–6.8.)

Chapter 6 Problems

6.1 *Engine Efficiencies*
(a) What is the maximum efficiency of a heat engine operating between reservoirs with temperatures of 20°C and 500°C? (S)
(b) Suppose an actual engine operating between these two temperatures produces 120 J of work and rejects 180 J of heat to the colder reservoir each cycle. What is the efficiency of this engine?

6.2 *Refrigerator.* A reversible refrigerator engine extracts heat from the inside of a refrigerator compartment kept at 8°C and rejects unwanted heat Q_H to its 20°C exterior. Find the work required to extract one calorie from the interior of the refrigerator compartment.

6.3 *Air Conditioner.* Recall from Section 5.5 that the coefficient of performance of a refrigerator operating between two heat reservoirs is Q_C/W.

(a) What is the coefficient of performance of an air conditioner that consumes 3 kW of electrical power and extracts heat from the house at a rate of 40 kW?

(b) What is the highest possible Q_C/W for an air conditioner operating between 25°C and 40°C?

6.4 *Three-Reservoir Heat Engine.* A reversible heat engine extracts heat $Q_H > 0$ from a reservoir at temperature T_H and heat $Q_M = aQ_H > 0$ from a reservoir at temperature $T_M \leq T_H$ while rejecting waste heat $Q_C > 0$ to a reservoir at temperature $T_C \leq T_M$.

(a) Derive an expression for the efficiency of this three-reservoir heat engine in terms of a and the three temperatures T_H, T_M, and T_C, where the efficiency is the total work produced divided by the total heat extracted from the two hotter reservoirs.

[Hint: Divide the reversible three-reservoir heat engine into two independent Carnot engines—one operating between reservoirs with temperatures T_H and T_M and the other operating between reservoirs with temperatures T_M and T_C.]

(b) Show that this expression reduces to expected results in the limits $a \to 0$ and $T_M \to T_C$.

6.5 *Newton's Law of Cooling.* A building is maintained at temperature T_H with a reversible heat pump operating between the building and a colder environment at temperature $T_C < T_H$. The heat pump consumes electrical power at a constant rate $\dot{W}$. The building also loses heat according to Newton's law of cooling, that is, at a rate $\alpha(T_H - T_C)$ where α is constant. Show that the building temperature is maintained at temperature

$$T_H = T_C + \frac{\dot{W}}{2\alpha}\left[1 + \sqrt{1 + \frac{4\alpha T_C}{\dot{W}}}\right].$$

6.6 *Radiation Cooling.* A Carnot engine generates work at a rate $\dot{W}$ by operating between reservoirs at temperatures T_C and $T_H > T_C$. The lower-temperature reservoir is a finite body with surface area A that maintains its temperature by radiating electromagnetic energy into space at a rate $\sigma_B A T_C^4$, where σ_B is a universal constant.

(a) Express $\dot{W}$ in terms of T_C, T_H, σ_B, and A.

(b) Suppose one wants to minimize the area A of the colder reservoir while generating work at a given rate $\dot{W}$ by extracting heat from a reservoir at given temperature T_H. What is this minimum area A and what is the temperature T_C of the colder reservoir that minimizes the area?

6.7 *Three-term Clausius Inequality.* An arbitrary—that is, not necessarily reversible—cycle extracts heat $Q_3 > 0$ from a reservoir of temperature T_3, extracts heat $Q_2 > 0$ from a reservoir of temperature T_2, and rejects heat $Q_1 > 0$ to a reservoir of temperature T_1. What is the relevant Clausius inequality?

6.8 *Finite Heat Reservoirs.* Two finite objects are identical in every way except that one has initial temperature T_C and the other initial temperature $T_H > T_C$. Assume that the objects neither expand nor contract upon changing temperature and that their heat capacity, C, is independent of temperature.

(a) Show that, according to calorimetric principles, when the objects are placed in direct thermal contact, their final equilibrium temperature is $(T_C + T_H)/2$.

(b) Suppose the two objects are brought into thermal equilibrium by producing work with a reversible heat engine—that is, a Carnot engine, operating between the two objects. Show that their final equilibrium temperature is $\sqrt{T_H T_C}$.

[Hint: Imagine heat exchanges of differential size.]

(c) How much work is produced in the process described in (b)?

Entropy

···

7.1 The Meaning of Reversibility

···

All quasistatic—that is, all indefinitely slow—heat transfers are necessarily reversible for the following reason. A quasistatic heat exchange between a system and its environment occurs only when the two are essentially in thermal equilibrium; otherwise the heat transfer is not indefinitely slow. Shifting the temperature of a system or its environment up or down an infinitesimal amount reverses a quasistatic heat transfer. Thus, reversible heat flows occur between objects at essentially the same temperature.

Quasistatic work, however, is not necessarily reversible; only work performed both quasistatically and without dissipation is completely reversible. Consider, for instance, the work performed on a fluid system contained within the Joule apparatus of Figure 2.2a. If the handle turns the paddles indefinitely slowly, the fluid occupies a series of equilibrium states as its temperature increases indefinitely slowly. Reversing the direction of the handle's turning requires a finite change in the applied torque. Furthermore, doing so would not reverse the order of equilibrium states through which the fluid passes. Stirring a fluid is not reversible because the work performed is dissipated via internal friction.

Sadi Carnot recognized a very practical consequence of reversibility in heat engines: reversible heat engines are the most efficient ones possible. Of course, the engineers of his day already knew that minimizing dissipation increased engine efficiency. But Carnot was the first to discover that minimizing temperature differences between the thermally conducting parts of an engine and its environment also increased efficiency. If engine efficiency is the goal, both sources of irreversibility—dissipation and finite-temperature heat-exchange—should be minimized.

Rudolph Clausius also applied the concept of reversibility—in his case, to advance the theory of thermodynamics. According to Clausius's theorem for reversible cycles,

$$\oint \frac{dQ_{rev}}{T} = 0, \tag{7.1}$$

where T is the common temperature of the system and the environment with which it exchanges heat. As we shall see, every kind of reversible interaction, work as well as heat, establishes a link between a system and its environment that irreversibility breaks. Here and elsewhere, I use the symbols dQ_{rev} and dW_{rev} (and Q_{rev} and W_{rev}) to emphasize that the indicated heat or work exchanged is reversible.

7.2 Entropy

Clausius's genius was to see that Equation (7.1) implies the existence of a new, purely thermodynamic state variable. Clausius called this new state variable *entropy* in parallel with the word *energy* and after the Greek η τροπη′ (*ē tropē*) or η στροφη′ (*ē strophē*), both of which refer to "turn"—appropriate since entropy regulates the turn or direction of thermodynamic processes. The Greek letter sigma, σ, beginning στροφη' may have provided Clausius with the conventional symbol, S, for entropy.

According to Clausius, a system changes its store of entropy by

$$dS \equiv \frac{dQ_{rev}}{T} \qquad (7.2)$$

during a differential reversible heat exchange. Recall that here T is the common temperature of the system and its environment and $dQ > 0$ ($dQ < 0$) when a system absorbs (rejects) heat.

However, we first need to prove that entropy exists and has the properties of a state variable. Because we want a coherent thermodynamics, we cannot simply assume that entropy with its usual properties is a state variable any more than we could (erroneously) assume that the heat absorbed by a system is a state variable. State variables are not simply words with arbitrarily assigned meanings. Empirical temperature is a meaningful state variable only because thermometers exist, and thermometers exist only because the zeroth law of thermodynamics is invariably observed. Internal energy is a meaningful state variable only because the first law of thermodynamics obtains. The proof that entropy is a state variable requires indirectly, through Clausius's theorem, both the first and second laws of thermodynamics.

Fortunately, the proof is quite simple. Consider a system that experiences a reversible cycle while exchanging heat with its environment. Its heat exchanges observe Clausius's theorem for reversible cycles [Eq. (7.1)]. Allow the cycle to begin in an initial state, called A; proceed reversibly through a continuous series of other states, denoted path 1, to an intermediate state B; and reversibly continue through other states, denoted path 2, to the initial state A at the end of the cycle. Figure 7.1 diagrams this cycle in a state space described by two generic variables, X and Y. Thus, the path integral in Equation (7.1) may be divided into two parts so that

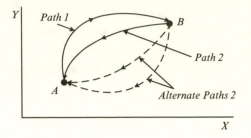

FIGURE 7.1 Reversible cyclic transformation of a system from state A to state B along path 1 and returning from state B to state A along path 2 or one of its alternatives. The variables X and Y describe a generic two-variable state space.

$$\int_{\substack{A \\ 1}}^{B} \frac{dQ_{rev}}{T} + \int_{\substack{B \\ 2}}^{A} \frac{dQ_{rev}}{T} = 0, \tag{7.3}$$

where paths 1 and 2 define each integral. Paths 1 and 2 are arbitrary, apart from representing reversible transformations between the two states. Because we can imagine fixing path 1 from A to B and replacing path 2 from B to A with alternative reversible paths 2, as illustrated in Figure 7.1,

$$\int_{B}^{A} \frac{dQ_{rev}}{T} = constant, \tag{7.4}$$

for any reversible process taking the system from state B to state A.

Since a path integral of the kind $\int_{B}^{A} dQ_{rev}/T$ can formally depend only on its endpoints or on the reversible path connecting its endpoints, and since the latter dependence is specifically denied by property (7.4), the integral $\int_{B}^{A} dQ_{rev}/T$ must depend only upon its endpoint states A and B. This conclusion justifies our defining an entropy function S of state A such that

$$S(A) = S(O) + \int_{O}^{A} \frac{dQ_{rev}}{T}, \tag{7.5}$$

where the reference state O and its entropy $S(O)$ are both, in principle, arbitrary parts of the definition. Typically, we assign the reference state and its entropy according to widely adopted convention or local convenience. Note that the SI unit of entropy is the joule per degree Kelvin (J/K).

The arbitrary element in definition (7.5) is completely analogous to the arbitrary element in the definition of energy. Neither entropy nor energy is known absolutely; each is known relative to an arbitrary value assigned to an arbitrary reference state. The reference state and its entropy are arbitrary because only differences in the entropies of two states are measurable. The measure of any entropy difference, say, $S(A) - S(B)$, is, according to definition (7.5) and the ordinary rules of integration,

$$S(A) - S(B) = \int_{B}^{A} \frac{dQ_{rev}}{T}. \tag{7.6}$$

When the two states, A and B, are differentially separated, (7.6) reduces to its differential equivalent,

$$dS = \frac{dQ_{rev}}{T}. \tag{7.7}$$

(See Problem 7.1.)

7.3 Entropy Generation in Irreversible Processes

Reversible processes are idealized processes because they unfold indefinitely slowly and without dissipation. All actual processes are irreversible and only approach reversibility in more or less degree. Clausius's theorem, $\oint dQ/T \leq 0$, shows how an actual irreversible process is limited by an ideal reversible one. Again, imagine a complete cycle divided into two parts: an irreversible process, Ir, taking the system from state A to state B and a reversible process, R, taking the system from state B back to state A. Since the cycle, considered as a whole, is irreversible, it must observe Clausius's theorem for irreversible cycles, which, in this case, is

$$\int\limits_{\substack{A \\ Ir}}^{B} \frac{dQ}{T} + \int\limits_{\substack{B \\ R}}^{A} \frac{dQ_{rev}}{T} < 0. \tag{7.8}$$

Since process R is reversible, we use (7.6) to replace $\int_{B_R}^{A} dQ_{rev}/T$ in (7.8) with $S(A) - S(B)$ so that

$$\int\limits_{\substack{A \\ IR}}^{B} \frac{dQ}{T} + S(A) - S(B) < 0, \tag{7.9}$$

that is,

$$S(B) - S(A) > \int\limits_{\substack{A \\ Ir}}^{B} \frac{dQ}{T}. \tag{7.10}$$

The following analysis reveals that an irreversible transformation of a thermally isolated system always increases a system's entropy. In thermal isolation the system exchanges no heat, that is, $dQ = 0$, and the right hand side of (7.10) vanishes. Consequently, $S(A) < S(B)$, or, in more suggestive notation,

$$S(f) > S(i), \tag{7.11}$$

where i denotes the initial state of a thermally isolated system and f its final state after irreversible transformation. Of course, rule (7.11) also describes irreversible transformations of systems that are completely, and not only thermally, isolated. (See Problems 7.2–7.4.)

7.4 The Entropy Generator

The device illustrated in Figure 7.2 consists of a Joule apparatus in thermal contact with a heat reservoir. As the mass, m, falls a distance h the paddle wheel turns and slowly stirs the fluid. During this

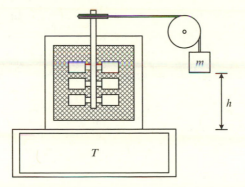

FIGURE 7.2 Entropy generator composed of Joule apparatus in thermal equilibrium with a heat reservoir. The mass falls, turns the paddle wheel, and stirs the fluid.

process the fluid remains in thermal equilibrium with the reservoir. The fluid maintains the same temperature, the same volume, and the same pressure, and, consequently, the same energy and entropy. Because work mgh is done on the fluid, a quantity of heat mgh must leave the fluid and enter the heat reservoir. Since the fluid and reservoir have the same temperature, T, the latter absorbs heat reversibly and increases its entropy by mgh/T. We call this device an *entropy generator*.

Let's apply the definition of entropy (7.6) and Clausius's theorem (7.10) to each part of the entropy generator, fluid and reservoir, and to its whole. As the weight falls, the fluid remains in the same state. Therefore,

$$S_{fluid}(f) - S_{fluid}(i) = 0. \tag{7.12}$$

However, the fluid rejects a quantity of heat, mgh. Applying (7.10) to the fluid produces the trivially correct inequality $-mgh/T < 0$. At the same time the heat reservoir reversibly absorbs heat mgh. Therefore, (7.6) applies to the heat reservoir and reduces to

$$S_{res}(f) - S_{res}(i) = mgh/T. \tag{7.13}$$

Finally, the composite system consisting of both the fluid and the heat reservoir is irreversibly transformed while thermally, but not mechanically, isolated from its environment. Thus, (7.10) produces

$$S_{total}(f) > S_{total}(i).$$ (7.14)

Equations (7.12)–(7.14), although having different origins, are clearly mutually consistent. (See Problem 7.5.)

7.5 Entropy Corollaries

Several corollaries follow from the laws of thermodynamics and the definition of entropy. Some have already been justified or illustrated; the remaining are easily developed.

Entropy is not conserved. Energy and entropy are similar kinds of quantities: both are extensive state variables, and neither is known absolutely but only in relation to a reference state. Yet there is an important difference between the two: total energy is always conserved; total entropy is never, in practice, conserved. The entropy generator, for instance, creates total entropy. According to (7.11), an irreversible process of a thermally isolated system creates entropy.

Entropy is an extensive state variable. An extensive state variable, such as energy or volume, is one whose quantity is directly proportional to the size of the system, while an intensive state variable, such as temperature or pressure, is independent of system size. Entropy is an extensive variable by virtue of its relation to other state variables. We know that the entropy of a system changes during a differential reversible heat exchange by $dS = dQ_{rev}/T$ and that, according to the first law,

$$dE = dQ_{rev} + dW.$$ (7.15)

Therefore,

$$dS = \frac{dE - dW}{T}.$$ (7.16)

Because the system energy E is an extensive variable, because the differential work dW done on the whole system is the sum of the work done on each of its parts, and because each part of an equilibrium system has a common temperature T, dS and its integral S are each the sum of quantities descriptive of parts of the system. For this reason, entropy is an extensive variable.

Occasionally, neither energy nor entropy appears to be proportional to the size of the system—as, for instance, when in dividing a system in two, new surfaces with non-negligible energy and entropy are created. But in such cases, conceiving of the surface itself as another system or phase in equilibrium with and open to the bulk phase preserves the fact that energy and entropy are extensive variables.

Entropy is additive. The entropy of two or more systems, even when these systems have different temperatures and/or structures, sum to the entropy of the whole conceived as a composite system. For instance, the entropy increase associated with the entropy generator is the sum of the changes in the entropy of the generator fluid and in the entropy of the generator reservoir. Thus, according to this corollary and equations (7.12)–(7.14),

$$S_{total}(f) - S_{total}(i) = [S_{fluid}(f) - S_{fluid}(i)] + [S_{res}(f) - S_{res}(i)]$$

$$= 0 + \frac{mgh}{T} > 0. \tag{7.17}$$

Conditions sufficient to establish this corollary follow from the facts that entropy is an extensive variable and that any two states can be connected with a reversible process. Thus, consider a composite system initially composed of two parts with a common temperature. Because entropy is an extensive variable, the entropy of the composite system $S(i)$ is the sum $S_1(i) + S_2(i)$ of the entropy of its two parts, that is,

$$S(i) = S_1(i) + S_2(i). \tag{7.18}$$

Furthermore, because any two states can be connected with a reversible process and because entropy change is calculable along a reversible process via $dS = dQ_{rev}/T$, we may conceive of a complex reversible process that takes the two systems 1 and 2 from each of their initial states with common temperature to two arbitrary final states with different temperatures. Suppose the net entropy increment of system 1 is $\Delta S_1 = S_1(f) - S_1(i)$ and the net entropy increment of system 2 is $\Delta S_2 = S_2(f) - S_2(i)$. Then we can, without fear of contradiction, offer as a definition that the net entropy increment of the composite system $\Delta S = S(f) - S(i)$ is

$$\Delta S \equiv \Delta S_1 + \Delta S_2. \tag{7.19}$$

Given Equation (7.18), Definition (7.19), and the meaning of ΔS_1, ΔS_2, and ΔS we find that

$$S(f) = S_1(f) + S_2(f). \tag{7.20}$$

Accordingly, the entropy of two or more arbitrary systems sums to the entropy of the composite system composed of both of the two arbitrary systems.

An increase in entropy degrades energy. In other words, irreversibility always diminishes the amount of work that can be extracted from an isolated system. Recall that the cyclic operation of a simple heat engine extracts heat $Q_H > 0$ from a heat reservoir at temperature T_H, produces work $W > 0$, and rejects waste heat $Q_C > 0$ to a colder heat reservoir at temperature $T_C < T_H$, as illustrated in Figure 7.3. (Here, we revert to our earlier convention according to which heat, Q, and work, W, are magnitudes.) Suppose this heat engine operates, as do all real engines, irreversibly. Since the composite system composed of the engine system and both heat reservoirs is isolated, its total entropy increases—that is, $\Delta S_{Ir} > 0$, where ΔS_{Ir} is the sum of the entropy change of each part of the composite system. Thus,

$$\Delta S_{Ir} = \Delta S_{engine} + \Delta S_H + \Delta S_C > 0. \tag{7.21}$$

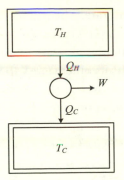

FIGURE 7.3 Simple heat engine that extracts heat $Q_H > 0$ from a reservoir at temperature T_H, produces work $W > 0$, and rejects waste heat $Q_C > 0$ to a colder reservoir at temperature $T_C < T_H$.

Given that $\Delta S_{engine} = 0$, $\Delta S_H = -Q_H/T_H$, and $\Delta S_C = Q_C/T_C$, Equation (7.21) becomes

$$\Delta S_{Ir} = \frac{-Q_H}{T_H} + \frac{Q_C}{T_C} > 0. \tag{7.22}$$

Using the first law of thermodynamics, $W_{Ir} = Q_H - Q_C$, to eliminate Q_C from (7.22) produces

$$T_C \Delta S_{Ir} = -W_{Ir} + Q_H\left(1 - \frac{T_C}{T_H}\right) > 0. \tag{7.23}$$

Clearly, $W_R = Q_H(1 - T_C/T_H)$ is the work that could be extracted from this system if the cycle were reversible. In these terms, the inequality of (7.23) reduces to

$$W_{Ir} < W_R. \tag{7.24}$$

The irreversible cycle produces less work from the same input heat Q_H than a reversible process would. Irreversibility degrades the energy into a form less available for producing work.

The entropy of a thermally isolated system cannot decrease. Only three possibilities exist: (1) A thermally isolated system remains in a particular equilibrium state. Then all its state variables, including entropy, remain constant. (2) A thermally isolated system transforms reversibly from one equilibrium state to another. Then its entropy remains constant even as its other state variables change. (3) A thermally isolated system transforms irreversibly from one equilibrium state to another. Then its entropy increases. Therefore, the entropy of a thermally isolated system never decreases. Of course, what holds for a thermally isolated system also holds for a completely isolated system. If the universe is an isolated system, the entropy of the universe cannot decrease.

A completely isolated system that can evolve only by diminishing its entropy is in stable equilibrium. A completely isolated system changes its state either reversibly or irreversibly. If reversibly, its entropy remains the same; if irreversibly, its entropy increases. Therefore, any process that demands that the entropy of an isolated system decrease is an impossible process. If an isolated system cannot change except by diminishing its entropy, that system cannot change. The equilibrium state of a system that admits no thermodynamic change is called a *stable equilibrium*. This corollary is also called the criterion of *maximum entropy* for the thermodynamic stability of a completely isolated system. (See Problems 7.6–7.7.)

7.6 Thermodynamic Arrow of Time

The dynamical laws describing gravitational, electromagnetic, and most nuclear particle interactions are reversible. Whatever evolves in one order can also evolve in reverse order. One has only to look at Newton's second law of motion for an example of this *time symmetry*. If a function of time, $f(t)$, solves Newton's second law, an identical function, $f(-t)$, with $-t$ replacing t, also solves Newton's second law.

The entropy corollaries highlight a fact that has no analog outside the realm of thermodynamics: thermodynamic processes are not time-reversible. Thermally isolated systems always evolve in the direction of increasing entropy—never in the direction of decreasing entropy. Entropy even quantifies time asymmetry. The larger the entropy of an isolated system, the more advanced in time its evolution.

Thermodynamic time asymmetry originates in the prohibitions of the second law of thermodynamics. The second law, for instance, allows work to be dissipated within a single-temperature heat reservoir but does not allow a single-temperature reservoir to produce work.

Sometimes one hears that only *natural* or *spontaneous* processes evolve in the direction of increasing entropy. Here, the words "natural" and "spontaneous" simply refer to changes in an isolated system. Systems that evolve without outside interference evolve spontaneously, however complicated or engineered their construction. Even processes maintained with external input must obey the second law of thermodynamics. When the context of any process is expanded to include all its interacting parts, the larger system always proceeds in the direction of increasing entropy.

Much effort has been directed to reducing thermodynamics, including the law of increasing entropy, to particle physics mediated through statistical methods. Interestingly, the qualitative difference between time-symmetric and time-asymmetric processes has meant that these efforts have not been entirely successful. The effort itself raises the question of what constitutes an explanation. Why, for instance, do we strive to explain some phenomena (for example, thermodynamic processes) in terms of others (for example, particle interactions)? Why are some laws considered more foundational than others? We cannot answer these questions here, but in my judgment the second law of thermodynamics remains a candidate for the foundational. Thermodynamics, alone among physical theories, is consistent with our sense of time and with a qualitative distinction between cause and effect, past and future. Thermodynamics alone provides a physical basis for what Arthur Eddington called the *arrow of time*.

. . . .

7.1 An object with heat capacity C absorbs heat and changes its temperature from T_i to T_f. What is the entropy increase?

7.2 *Entropy Change, I.* Does the entropy of the following system increase, decrease, or remain the same as it experiences each of the following changes of state?

(a) One gram of water absorbs enough heat at 373 K and atmospheric pressure to evaporate. (**S**)

(b) One gram of water freezes at standard temperature and pressure.

(c) A 1-kg bag of sand falls from a second-story window and lands on the pavement below.

(d) A frictionless piston very slowly compresses one mole of gas contained within an adiabatic jacket.

(e) One mole of gas is reversibly and isothermally compressed.

(f) A closed stopcock keeps one mole of gas in one part of a glass container. The other part is evacuated. The stopcock is opened. See Figure 7.4.

(g) The composite system is the hot coffee contained within an insulated thermos and an ice cube. The ice cube is dropped into the thermos and the thermos lid is replaced. The ice cube melts.

7.3 *Entropy Change, II.* Calculate the entropy change of these systems as a result of the following processes. (When necessary, use the data supplied in Tables 3.1 and 3.2.) Express all answers in SI units.

(a) Twenty-five grams of aluminum melts. (**S**)

(b) Ten grams of steam at 100°C and atmospheric pressure condense to liquid water at the same temperature and pressure.

(c) One kilogram of hot (80.0°C) aluminum is placed into 2 L of 20.0°C water in a container with negligible heat capacity and adia-

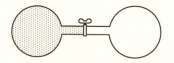

FIGURE 7.4   A closed stopcock keeps one mole of gas in one side of the container. The other side is evacuated. (Used in Problem 7.2f.)

batic walls. The water and aluminum reach thermal equilibrium. Assume the specific heats of aluminum and water are constants. [Hint: First find the temperature of the final equilibrium state. Then imagine a reversible heat flow process that takes the aluminum and the water from their initial state to their final state and calculate the associated entropy change. Since entropy is a state variable, the entropy increment of the imagined reversible process must be the same as the entropy increment of the actual irreversible process.]

7.4 *Carnot Cycle.* Recall that a Carnot cycle is a reversible cycle that extracts heat Q_H from a hot reservoir at temperature T_H, produces work W, and rejects waste heat Q_C to a cold reservoir at temperature $T_C < T_H$. Carnot cycles are, necessarily, composed of four distinct reversible parts: two isothermal and two *isentropic* (constant entropy) processes. Because the heat reservoirs are single-temperature and heat is exchanged reversibly, heat exchange occurs only in the isothermal processes. Since $dS = dQ_{rev}/T$, the reversible isentropic processes are, necessarily, adiabatic. Express the following in terms of the heat reservoir temperatures, T_H and T_C, and the entropies of the system, S_1, and $S_2 > S_1$, during each adiabatic process.

(a) The heat absorbed during the isothermal process $(S_1, T_H) \rightarrow (S_2, T_H)$ (S)

(b) The heat rejected during the isothermal process $(S_2, T_C) \rightarrow (S_1, T_C)$

(c) The total work produced in one cycle

(d) The total area contained by the cycle in $S - T$ space

[Hint: You may want to diagram the Carnot cycle in $S - T$ space.]

7.5 *Electric Current.* An electric current of 200 mA is maintained for 3 sec in a 20-ohm resistor. The temperature of the resistor remains constant at 25°C.

(a) What is the change in the entropy of the resistor?

(b) What is the change in the entropy of the resistor's environment?

[Hint: Consider the similarity with the entropy generator of Section 7.4.]

7.6 *Equilibration.* Consider two identical, finite blocks of metal each with constant heat capacity C. Initially, one is at temperature T_H and the other is colder, with temperature $T_C < T_H$. [Hint: Review Problem 6.8]

(a) The two blocks are placed in thermal contact and allowed to equili-
brate. According to Problem 6.8 their final temperature is $(T_H + T_C)/2$. What is the total change in the entropy of the two blocks?

(b) Imagine the two blocks, again at temperatures T_H and $T_C < T_H$, are
allowed to equilibrate by means of a reversible heat engine operating between the two blocks. According to Problem 6.8 their final
temperature is $\sqrt{T_H T_C}$. What is the total change in the entropy of
the two blocks?

7.7 *Benzene*. A flask containing one mole of liquid benzene at its normal freezing point of 5.5°C is brought into thermal contact with a large
ice-water reservoir until the benzene has frozen solid. The benzene remains
at 5.5°C. The molecular weight of benzene is 78.1 g/mole, its heat of fusion
is 30.3 cal/g, and its specific heat is 0.416 cal/(g °C).

(a) What is the decrease of the entropy of the benzene?

(b) What is the total increase of the entropy of the combined benzene
and ice-water reservoir system?

Fluid Variables

....

8.1 What Is a Fluid?

....

The word *fluid* suggests a material that takes the shape of its container. A gas is a compressible fluid, a liquid is a relatively incompressible fluid, and a solid, because of its rigidity, is not a fluid at all. But here we adopt another, closely related but distinct use of the word *fluid*. Any system adequately described by the variables pressure, P, volume, V, and temperature, T, is a fluid. In principle, a fluid can realize values of P, V, and T appropriate to a material in any of its phases—gas, liquid, or solid. Of course, a fluid model of a solid ignores its rigidity—and rigidity distinguishes solids from liquids. But the function of any model is to abstract certain features from a too-complex reality and ignore others.

Useful fluid variables neatly divide themselves into two kinds: extensive and intensive. Extensive variables, such as system mole number, n; volume, V; internal energy, E; and heat capacity, C, are directly proportional to the size of the system and, consequently, in any one system directly proportional to each other. Intensive variables, such as temperature, T; pressure, P; and molar density, n/V, are independent of system size. Grasping this distinction, illustrated in Figure 8.1, allows one to recognize which relations among fluid variables are quantitatively coherent and which are not.

Fluid descriptions are appropriate when pressure is single-valued. Figure 8.2 shows a pressure-measuring device in different positions and orientations. The device consists of a spring attached to a plate. The material whose pressure is to be measured is in front of the plate, completely separated from the evacuated region behind the plate. The compression (or extension) of the spring, relative to its position when a vacuum is on both sides of the plate, measures the material pressure. When the compression or extension is independent of the orientation of the measuring device, the material pressure is *isotropic*. When the compression or extension is independent of the position of the measuring device, the material pressure is *homogeneous*. When the system pressure is both isotropic and homogeneous, a

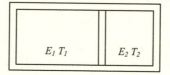

FIGURE 8.1   One fluid system divided into two parts. Since the energy is an extensive state variable, the energy of the whole is $E = E_1 + E_2$. Since temperature is an intensive state variable, the temperature of the whole is $T = T_1 = T_2$.

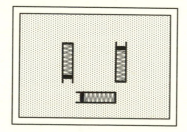

FIGURE 8.2   Pressure-measuring devices. Their orientation and position test whether the material pressure is isotropic and homogeneous.

single pressure, P, adequately describes the system. Solids typically have different pressures in different directions. For this reason, solid pressures are best represented by a *stress tensor*—an important tool but one that lies beyond the scope of this book.

The SI unit of pressure is the Pascal (Pa). By definition 1 Pa = 1 N/m^2. However, other pressure units are also common. Relatively high pressures are sometimes measured in bars or atmospheres where 1 bar = 10^5 Pa and 1 atm = 1.01×10^5 Pa = 1.01 bar. Low pressures—for instance, those achieved in vacuum systems—are often measured in *torrs, where* 1 torr = 1/760 atm. These and other pressure units are listed in Appendix A. When part of a system pulls on its neighboring parts, its pressure is negative. Solids can pull as well as push, but so can liquids. The maximum negative pressure sustainable by a material is called its *ultimate tensile strength*. Steel, for instance, has an ultimate strength close to 10^9 Pa, while liquids have ultimate strengths on the order of 10^6 Pa. (See Problems 8.1–8.2.)

8.2 Reversible Work

Reversible interactions link a system to its environment. For instance, reversible heat exchanges proceed in increments dQ_{rev} that are related to system state variables and their increments by $dQ_{rev} = TdS$. Reversible work also proceeds in increments dW_{rev} that are likewise related to system state variables and their increments. When the system is a fluid, this relation is quite transparent.

Figure 8.3 illustrates reversible work preformed on a fluid system. Here, a piston quasistatically and without dissipation—that is, reversibly—compresses a fluid. The force exerted by the piston head on the fluid is only infinitesimally larger than the force exerted by the fluid on the piston head; otherwise, the compression would be too fast and, consequently, irreversible. The work W_{rev} performed by the environment on the fluid in reversibly compressing the fluid through a distance Δx is

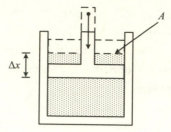

FIGURE 8.3 Compression of a fluid system. When the compression is reversible, the work done on the fluid is $\Delta W_{rev} = -P \cdot \Delta V$, where $\Delta V = -A \cdot \Delta x$.

$$W_{rev} = PA\Delta x$$
$$= -P\Delta V, \tag{8.1}$$

where A is the surface area of the piston head, P is the fluid pressure, and V the fluid volume. A minus sign appears in Equation (8.1) because $\Delta V < 0$ when $W > 0$. Consequently, infinitesimal reversible work dW_{rev} is related to infinitesimal compression dV and fluid pressure P by

$$dW_{rev} = -PdV. \tag{8.2}$$

Reversible work of any kind can always be expressed in terms of system state variables and changes in state variables. Irreversible work cannot be so expressed.

The equation $dW_{rev} = -PdV$ does not imply that W_{rev} is a state variable for the same reason that the equation $dQ_{rev} = TdS$ does not imply that Q_{rev} is a state variable. The amount of work performed, W_{rev}, and the quantity of heat exchanged, Q_{rev}, in a reversible process depends upon the path chosen to connect its initial and final states. Many texts invent a special notation—for example, $đW$ in place of dW—to distinguish "imperfect" differentials whose integrals are path-dependent from "perfect" differentials of state variables whose integrals depend only upon the limits of integration. Here we forgo any special notation and simply recall that W, W_{rev}, Q, and Q_{rev} are not state variables; they quantify processes rather than states. (See Problems 8.3–8.4.)

The first law of thermodynamics for reversible processes, $dE = dQ_{rev} + dW_{rev}$; the second law of thermodynamics implicit in $dQ_{rev} = TdS$; and the expression for reversible work on fluid systems, $dW_{rev} = -PdV$, together imply a differential relation

$$dE = TdS - PdV \qquad (8.3)$$

among the fluid variables E, T, S, P, and V. While Equation (8.3) is motivated by a reversible process, the result describes, not necessarily a process, but rather, a constraint among fluid variables imposed by the laws of thermodynamics. We call (8.3) the *fundamental constraint* on fluid systems. The fundamental constraint (8.3) is a treasure trove of information on the relations among fluid variables.

The mathematical structure of the fundamental constraint $dE = TdS - PdV$ is identical to the structure assumed by the differential of a function $x(y,z)$ of two independent variables y and z, that is, by

$$dx = \left(\frac{\partial x}{\partial y}\right)_z dy + \left(\frac{\partial x}{\partial z}\right)_y dz. \qquad (8.4)$$

Important consequences follow immediately from this structure. But first note the possibly unusual practice of attaching subscripts to the partial derivatives $(\partial x/\partial y)_z$ and $(\partial x/\partial y)_y$ in (8.4). When the identity of and distinction between dependent, x, and independent, y and z, variables is obvious, such notation is unnecessary. But fluid systems are described with a large number of variables—for example, at least P, V, T, E, and S—each pair of which can serve as independent variables and each one of which can be an independent variable. Not to clearly identify the independent variables risks missing the crucial difference between, say, $(\partial E/\partial V)_S$ and $(\partial E/\partial V)_T$.

Because the mathematical identity (8.4) takes the same form as the fundamental constraint (8.3), the terms in (8.3) and (8.4) must enter into analogous relations. In particular, because the variables x, $(\partial x/\partial y)_z$, y, $(\partial x/\partial y)_y$, and z occupy the same position in (8.4) as the fluid variables E, T, S, $-P$, and V occupy in (8.3), it follows that

$$T = \left(\frac{\partial E}{\partial S}\right)_V \tag{8.5}$$

and

$$P = -\left(\frac{\partial E}{\partial V}\right)_S . \tag{8.6}$$

Equations (8.5) and (8.6) are called *equations of state*. Furthermore, because the order of partial differentiation is inconsequential, that is, because $\partial^2 E/\partial S\partial V = \partial^2 E/\partial V\partial S$, such *cross-differentiation* applied to these equations of state, (8.5) and (8.6), implies

$$\left(\frac{\partial T}{\partial V}\right)_S = -\left(\frac{\partial P}{\partial S}\right)_V , \tag{8.7}$$

one of four possible fluid *Maxwell relations,* so called after James Clerk Maxwell (1831–1879), who first collected them.

The function $E(S,V)$ and the fundamental constraint $dE = TdS - PdV$ completely describe a fluid. For this reason $E(S,V)$ is called a thermodynamic *characterizing function*. After all, its derivatives $(\partial E/\partial S)_V$ and $(\partial E/\partial V)_S$ generate expressions, through (8.5) and (8.6), for the remaining fluid variables, respectively, T and P.

Of course, the fundamental constraint $dE = TdS - PdV$ might, equivalently, be written as $dS = dE/T + (P/T)dV$ or as $dV = (T/P)dS - dE/P$. These, in turn, imply the existence of other characterizing functions $S(E,V)$ and $V(S,E)$ with their own proper independent variables and consequences parallel in form to (8.5) and (8.6), for instance, the equations of state $1/T = (\partial S/\partial E)_V$ and $P/T = (\partial S/\partial V)_E$. Such expressions are often very useful.

Other systems that allow different kinds of reversible work have their own fundamental constraints. A form that allows for different kinds of reversible work is

$$dW_{rev} = \sum_i F_i dX_i , \tag{8.8}$$

where the F_i and X_i are known, respectively, as *generalized forces* and their associated *generalized displacements*. The generalized fundamental constraint is

$$dE = TdS + \sum_i F_i dX_i. \tag{8.9}$$

A fluid, of course, admits only one kind of reversible work: one for which the generalized force F is $-P$ and the generalized displacement X is V. In this case, the generalized fundamental constraint (8.9) reduces to the particular fundamental constraint (8.3) for fluids. (See Problems 8.5–8.6.)

8.4 Enthalpy

The energy characterizing function $E(S,V)$ is not always convenient. Because its independent variables, S and V, are not so easily manipulated or measured, other extensive state functions with the dimension of energy and independent variables from among the possibilities P, V, T, and S have been defined. The *enthalpy*

$$H \equiv E + PV \tag{8.10}$$

is one; we will encounter others in Section 8.5. Formally (8.10) defines a *Legendre transformation*. The goal of a Legendre transformation is to replace the dependent variable $E(S,V)$ with another, say $H(S,P)$, having independent variables, S and P in place of S and V. The enthalpy's independent variables are more convenient in special circumstances.

The effect of the Legendre transformation (8.10) becomes clear when we use it to replace E with H in the fundamental constraint $dE = TdS - PdV$. For instance, from the definition $H \equiv E + PV$ we have $dH = dE + PdV + VdP$. Using this result to eliminate dE in $dE = TdS - PdV$ produces

$$dH = TdS + VdP. \tag{8.11}$$

Thus, the enthalpy has its own proper independent variables, S and P, and its own equations of state,

$$T = \left(\frac{\partial H}{\partial S}\right)_P \qquad (8.12)$$

and

$$V = \left(\frac{\partial H}{\partial P}\right)_S. \qquad (8.13)$$

These, in turn, imply the cross-differentiation

$$\left(\frac{\partial T}{\partial P}\right)_S = \left(\frac{\partial V}{\partial S}\right)_P, \qquad (8.14)$$

which is another Maxwell relation.

Enthalpy simplifies the description of a constant pressure process such as might unfold in a system exposed to the atmosphere. In a reversible *isobaric*—that is, constant pressure—process the enthalpy version of the fundamental constraint (8.11) reduces to

$$dH = TdS$$
$$= dQ_{rev}. \qquad (8.15)$$

Therefore,

$$\Delta H = Q_{rev}. \qquad (8.16)$$

In this special case the change in enthalpy, ΔH, is identical to the heat, Q_{rev}, reversibly absorbed or rejected by the fluid. For this reason, enthalpy is sometimes referred to as *heat content*—a designation we should avoid. Heat, after all, is never contained within a system; heat is not a state variable.

Heat exchange Q also equals enthalpy change ΔH in many irreversible processes. Consider, for instance, the special but frequently realized case of chemicals rapidly reacting while immersed in a fluid system that is open to the atmosphere. Suppose both the

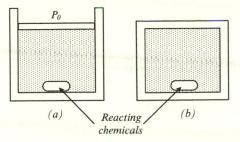

(a)   Reacting   (b)
      chemicals

FIGURE 8.4   Reacting chemicals surrounded by a fluid contained in a calorimeter with adiabatic boundaries. (a) Constant pressure calorimeter with
$Q = \Delta H$. (b) Constant volume bomb calorimeter with $Q = \Delta E$.

atmospheric pressure exerted on the system and the system volume
remain well-defined state variables as the reaction proceeds (Fig.
8.4a). In this case the work W performed by the atmosphere on
the fluid is $-P_o(V_f - V_i)$, where P_o is atmospheric pressure and $V_f - V_i$
is the change in the system volume. Consequently, the first law of
thermodynamics applied to the fluid becomes

$$E_f - E_i = Q - P_o(V_f - V_i), \tag{8.17}$$

which on rearranging asserts that

$$Q = (E_f + P_o V_f) - (E_i + P_o V_i)$$

$$= H_f - H_i \tag{8.18}$$

$$= \Delta H.$$

This step requires that the fluid return to its initial mechanical equilibrium with $P = P_o$ after the chemical reaction is complete.

Figure 8.4b illustrates another circumstance in which irreversible heat exchange equals the change in an energy function. Here,
the fluid volume is held constant, as in a so-called *bomb calorimeter*,
while a chemical reaction proceeds irreversibly. Therefore, $W = 0$
and the first law reduces to $\Delta E = Q$. (See Problem 8.7.)

8.5 Helmholtz and Gibbs Free Energies

Other thermodynamic characterizing energy functions are similarly useful in different but parallel circumstances. The *Helmholtz free energy*

$$A \equiv E - TS \tag{8.19}$$

and the *Gibbs free energy*

$$G \equiv E + PV - TS \tag{8.20}$$

also compose Legendre transformations that transform the fundamental constraint $dE = TdS - PdV$ into its equivalents

$$dA = -SdT - PdV \tag{8.21}$$

and

$$dG = -SdT + VdP. \tag{8.22}$$

Thus, $A = A(T,V)$ and $G = G(T,P)$. The Helmholtz and Gibbs forms of the fundamental relation, (8.21) and (8.22), imply a pair of equations of state, and each pair of these produces, via cross-differentiation, another Maxwell relation. The four Maxwell relations generated in this way are

$$\left(\frac{\partial T}{\partial V}\right)_S = -\left(\frac{\partial P}{\partial S}\right)_V, \tag{8.23}$$

$$\left(\frac{\partial T}{\partial P}\right)_S = \left(\frac{\partial V}{\partial S}\right)_P, \tag{8.24}$$

$$\left(\frac{\partial S}{\partial V}\right)_T = \left(\frac{\partial P}{\partial T}\right)_V, \tag{8.25}$$

and

$$\left(\frac{\partial S}{\partial P}\right)_T = -\left(\frac{\partial V}{\partial T}\right)_P. \tag{8.26}$$

The Helmholtz free energy is convenient when temperature and volume are easily controlled or measured as, say, when a fluid slowly evolves in thermal equilibrium with a constant temperature environment. When such a process is also reversible, the fundamental constraint $dA = -SdT - PdV$ reduces to $dA = -PdV$, which, in turn, integrates to

$$\Delta A = -\int_{V_i}^{V_f} PdV. \tag{8.27}$$

Thus, reversible work performed on, or by, an isothermal fluid increases, or decreases, its Helmholtz free energy. Its symbol, A, derives from the first letter of the German word *Arbeit,* meaning "work." Another important use of $A(T,V)$ is to relate statistical models to their correlative thermodynamic descriptions. But this application lies beyond the boundary of this book.

The outstanding application of the Gibbs free energy, $G(T,P)$, is to a system experiencing a phase transition while in thermal and mechanical contact with its environment. Of course, a system experiencing a reversible, isobaric, and isothermal process conserves its Gibbs free energy. We take up this important application in Chapter 12. (See Problem 8.8.)

8.6 Partial Derivative Rules

At a certain point the study of thermodynamics appears to degenerate into manipulating a mass of partial derivatives. We have already reached this point, but the charge of "degeneration" is unfair. After all, partial derivatives expressing how one variable depends on another when a third variable is held constant are typical experimental outcomes. To understand how these outcomes imply yet other derivatives and functions not so easily measured would seem a great advance. The tools required for these implications are the theorems or rules that follow from the general properties of state functions.

Mastering these rules allows us to push through any mass of partial derivatives.

Recall that every state function $x = x(y,z)$ with two independent variables implies the differential form

$$dx = \left(\frac{\partial x}{\partial y}\right)_z dy + \left(\frac{\partial x}{\partial z}\right)_y dz. \tag{8.28}$$

Alternatively, we might solve $x = x(y,z)$ for $y = y(x,z)$ and form

$$dy = \left(\frac{\partial y}{\partial x}\right)_z dx + \left(\frac{\partial y}{\partial z}\right)_x dz. \tag{8.29}$$

Among the three differentials dx, dy, and dz only two can be independent. Eliminating dy from among (8.28) and (8.29) produces the relationship

$$\left[\left(\frac{\partial x}{\partial y}\right)_z \left(\frac{\partial y}{\partial x}\right)_z - 1\right] dx + \left[\left(\frac{\partial x}{\partial y}\right)_z \left(\frac{\partial y}{\partial z}\right)_x + \left(\frac{\partial x}{\partial z}\right)_y\right] dz = 0 \tag{8.30}$$

between two independent differentials dx and dz. Choosing $dz = 0$ generates the *reciprocal rule*,

$$\left(\frac{\partial x}{\partial y}\right)_z = \frac{1}{\left(\dfrac{\partial y}{\partial x}\right)_z}, \tag{8.31}$$

which replaces a partial derivative with the reciprocal of its inverse. Choosing $dx = 0$ generates

$$\left(\frac{\partial x}{\partial y}\right)_z \left(\frac{\partial y}{\partial z}\right)_x = -\left(\frac{\partial x}{\partial z}\right)_y, \tag{8.32}$$

which, given the reciprocal rule, is equivalent to the *reciprocity rule*

$$\left(\frac{\partial x}{\partial y}\right)_z \left(\frac{\partial y}{\partial z}\right)_x \left(\frac{\partial z}{\partial x}\right)_y = -1. \tag{8.33}$$

The reciprocity rule generates new derivatives from old ones by cycling the dependent and independent variables into new positions. The pattern of the variable position in (8.33) is easily remembered.

We will also exploit the *chain rule*. Recall, from single-variable calculus, that if $x = x(y)$ and $y = y(z)$, then $x = x[y(z)]$ and

$$\frac{dx}{dz} = \left(\frac{dx}{dy}\right)\left(\frac{dy}{dz}\right). \tag{8.34}$$

Suppose, though, that each dependent variable is a function of two independent variables so that $x = x(y,w)$ and $y = y(z,w)$. Then $x = x[y(z,w),w]$. Forming the partial derivatives $(\partial x/\partial z)_w$ and $(\partial x/\partial w)_z$ produces

$$\left(\frac{\partial x}{\partial z}\right)_w = \left(\frac{\partial x}{\partial y}\right)_w \left(\frac{\partial y}{\partial z}\right)_w \tag{8.35}$$

and

$$\left(\frac{\partial x}{\partial w}\right)_z = \left(\frac{\partial x}{\partial y}\right)_w \left(\frac{\partial y}{\partial w}\right)_z + \left(\frac{\partial x}{\partial w}\right)_y, \tag{8.36}$$

both of which are applications of the chain rule. We refer to the latter (8.36) as the *multivariate chain rule*.

All partial derivative identities involving fluid variables can be derived from the reciprocal rule, the reciprocity rule, a chain rule, and a fundamental constraint. For example, a single fundamental constraint directly implies only a single Maxwell relation. But the three rules—reciprocal, reciprocity, and chain—can transform any one fluid Maxwell relation into the three others. As an example, apply reciprocity to either side of $(\partial T/\partial V)_S = -(\partial P/\partial S)_V$. Then apply the reciprocal and the chain rule to the result. This procedure generates either $(\partial T/\partial P)_S = (\partial V/\partial S)_P$ or $(\partial T/\partial P)_V = (\partial V/\partial S)_T$. Another application of the same procedure to either of these produces the fourth Maxwell relation $(\partial S/\partial P)_T = -(\partial V/\partial T)_P$. (See Problems 8.9–8.10.)

8.7 Thermodynamic Coefficients

Of all possible partial derivatives of the form $(\partial X/\partial Y)_Z$ where X, Y, and Z are any three of the five fluid variables E, T, S, P, and V, those partial derivatives with easily manipulated, controlled, or measured variables are the most useful. Some of these have been normalized and given special names. The isothermal compressibility

$$\kappa_T \equiv \frac{-1}{V}\left(\frac{\partial V}{\partial P}\right)_T, \tag{8.37}$$

its inverse the bulk modulus

$$B \equiv -V\left(\frac{\partial P}{\partial V}\right)_T, \tag{8.38}$$

and the adiabatic compressibility

$$\kappa_S \equiv \frac{-1}{V}\left(\frac{\partial V}{\partial P}\right)_S \tag{8.39}$$

quantify the degree to which a fluid system can, under given conditions, be compressed. The larger the compressibility, the more easily compressed the system. The isobaric expansivity

$$\alpha_P \equiv \frac{1}{V}\left(\frac{\partial V}{\partial T}\right)_P \tag{8.40}$$

quantifies the degree to which a system expands with increase in temperature. The isochoric pressure coefficient

$$\alpha_V \equiv \frac{1}{P}\left(\frac{\partial P}{\partial T}\right)_V \tag{8.41}$$

can be measured or, alternatively, inferred from an identity, $\alpha_V = \alpha_P/(P\kappa_T)$, that follows from the reciprocity and reciprocal rules applied to the variables P, V, and T. Each of the so-called *thermodynamic coefficients* κ_T, κ_S, α_P, and α_V is the inverse of an intensive variable that is itself a function of state. Their values at the material's *triple point*, where gas, liquid, and solid phases coexist, and at other standard states are available in reference books. (See Problem 8.11.)

8.8 Heat Capacities

A fluid has at least two heat capacities that are distinguished from one another by the conditions maintained as the fluid exchanges heat. The heat capacity with volume held constant is expressed by

$$
\begin{aligned}
C_V &\equiv \left(\frac{dQ_{rev}}{dT}\right)_V = T\left(\frac{\partial S}{\partial T}\right)_V \\
&= \left(\frac{\partial E}{\partial T}\right)_V,
\end{aligned}
\tag{8.42}
$$

where the last step follows from inspecting the fundamental relation $dE = TdS - PdV$. In contrast, the heat capacity with pressure held constant takes the form

$$
\begin{aligned}
C_P &\equiv \left(\frac{\partial Q_{rev}}{\partial T}\right)_P = T\left(\frac{\partial S}{\partial T}\right)_P \\
&= \left(\frac{\partial H}{\partial T}\right)_P,
\end{aligned}
\tag{8.43}
$$

where the last step follows from the enthalpy form of the fundamental relation $dH = TdS + VdP$.

Typically, $C_P > C_V$ because more heat is required to increase the temperature of a fluid when it is free to expand and perform work on its environment than when the fluid is not free to expand. Analytically, the inequality $C_P > C_V$ follows from the equations $C_P = T(\partial S/\partial T)_P$ and $C_V = T(\partial S/\partial T)_V$ and the multivariate chain rule applied to the function form $S(T,P) = S[T,P(V,T)]$, that is,

$$\left(\frac{\partial S}{\partial T}\right)_V = \left(\frac{\partial S}{\partial T}\right)_P + \left(\frac{\partial S}{\partial P}\right)_T \left(\frac{\partial P}{\partial T}\right)_V. \tag{8.44}$$

Therefore,

$$C_P = C_V - T\left(\frac{\partial S}{\partial P}\right)_T \left(\frac{\partial P}{\partial T}\right)_V. \tag{8.45}$$

The Maxwell relation $(\partial S/\partial P)_T = -(\partial V/\partial T)_P$ transforms the latter into

$$C_P - C_V = T\left(\frac{\partial V}{\partial T}\right)_P \left(\frac{\partial P}{\partial T}\right)_V. \tag{8.46}$$

Finally, the reciprocity and reciprocal rules applied to $(\partial P/\partial T)_V$ transform (8.46) into

$$C_P - C_V = -T\left(\frac{\partial V}{\partial T}\right)_P^2 \left(\frac{\partial P}{\partial V}\right)_T. \tag{8.47}$$

Therefore, $C_P > C_V$ as long as $(\partial P/\partial V)_T < 0$. The latter, as we will find in Chapter 11, is a requirement for the *intrinsic stability* of a fluid system. (See Problems 8.12–8.18.)

Chapter 8 Problems

8.1 *Pressure Units.* Complete the following table.

	Pascal	atm	torr	mm Hg
(a)	50×10^3			
(b)		1		
(c)			10^{-6}	
(d)				30

8.2 *Extensive and Intensive Fluid Variables.* The internal energy E of n moles of a particular fluid system is related to its pressure P and volume V by

$$E = aPV^2,$$

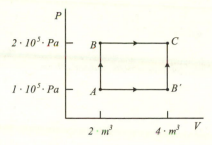

FIGURE 8.5 Two different reversible paths, ABC and AB′C, take a system from state A to state C. (Used in Problem 8.3)

where $a = 2.5 \times 10^{-5}/\mathrm{m}^3$ and $n = 5$. The system's internal energy, E, volume, V, and mole number, n, are extensive variables, while the system's pressure, P, is an intensive variable. Determine the complete dependence of E on P, V, and n by determining the dependence of a on n.

8.3 *Reversible Paths*. A fluid system is reversibly transformed from state A to state C along the two paths shown in Figure 8.5. In answering questions (a)–(e) one does not need to know the system's equations of state.

(a) Along which path, ABC or AB′C, is more work done by the system?

(b) Along which path is more heat absorbed by the system?

(c) Suppose the system is carried through a complete cycle ABCB′A. How much work is done by the system? How much heat is absorbed by the system?

(d) What is the net change in the system energy when carried through ABCB′A?

(e) What is the net change in the system entropy when carried through ABCB′A?

8.4 *Reversible Work*. A syringe closed at one end is composed of a piston and cylinder and contains a fluid system. Suppose the piston, whose cross-sectional diameter is 3.5 cm², is initially 7.0 cm from the bottom of the cylinder. The syringe piston moves without friction and does 0.50 J of work while slowly compressing the fluid 3.0 mm. (Assume the fluid pressure, P, is changed very little during this compression.) Find P in atmospheres.

8.5 *Hypothetical Characterizing Function.* Suppose the characterizing function of a fluid takes the form

$$E(S,V) = a\frac{S^2}{V},$$

where a is a constant that characterizes the particular hypothetical system.
   (a) Determine the equations of state that assume the form $T = T(S,V)$
       and $P = P(S,V)$. (S)
   (b) Show that, as expected, the Maxwell relation $(\partial T/\partial V)_S = -(\partial P/\partial S)_V$
       reduces to a tautology.

8.6 *Entropy Form of the Fundamental Relation.* We rewrite the fundamental constraint $dE = TdS - PdV$ in its equivalent entropy form $dS = (1/T)dE + (P/T)dV$.
   (a) Find the forms of the two equations of state that follow from the
       entropy form of the fundamental constraint—that is, find the derivatives of the dependent variable S in terms of its proper independent variables E and V that give expressions for T and P.
   (b) Use cross-differentiation to find the relation between the derivatives
       of T and P.

8.7 *Chemical Reaction.* A mole of the solid-phase chemical $MgCO_3$ decomposes into a mole of gas CO_2 and a mole of solid MgO at a constant temperature of 900 K and atmospheric pressure by absorbing 26,000 cal of heat. Both initial and final states may be described with fluid variables. The molar volumes of $MgCO_3$, MgO, and CO_2 are, respectively, 0.028 L, 0.011 L, and 740 L. Find
   (a) $\Delta H = H_f - H_i$, and
   (b) $\Delta E = E_f - E_i$.

8.8 *Maxwell Relations, I.* Derive the Maxwell relations from the fundamental constraint $dE = TdS - PdV$, the definitions $H = E + PV$, $A = E - TS$, and $G = E + PV - TS$, and cross-differentiation.

8.9 *Maxwell Relations, II.* Derive the following Maxwell relations from $(\partial T/\partial V)_S = -(\partial P/\partial S)_V$ and the reciprocal, reciprocity, and chain rules. List each rule as it is invoked.

(a) $(\partial T/\partial P)_S = (\partial V/\partial S)_P$

(b) $(\partial T/\partial V)_P = -(\partial P/\partial S)_T$

(c) $(\partial T/\partial P)_V = (\partial V/\partial S)_T$

8.10 *Reciprocity.* The equation $2x^2 + y^3z^4 = 0$ defines a relation among the three variables x, y, and z. Calculate each partial derivative $(\partial x/\partial y)_z$, $(\partial y/\partial z)_x$, and $(\partial z/\partial x)_y$ independently by taking the appropriate partial derivatives of the equation. Form the product $(\partial x/\partial y)_z(\partial y/\partial z)_x(\partial z/\partial x)_y$ and show that, as expected, it equals -1.

8.11 *Thermodynamic Identity, I.* Derive the identity $\alpha_V = \alpha_P/P\kappa_T$ from the reciprocity rule applied to the fluid variables.

8.12 *Helmholtz Free Energy.* Determine the following thermodynamic properties of a system whose Helmholtz free energy, $A(T,V)$, is known. Each property should be expressed in terms of A, its derivatives, and the independent variables T and V. Use the Maxwell relations along with the reciprocal, reciprocity, and chain rules when necessary.

(a) P, (b) S, (c) E, (d) C_V, (e) κ_T, (f) α_P, and (g) α_V.

8.13 *Heat Capacities.* Starting from definitions of C_P and C_V, show that $C_P - C_V = TV\alpha_P^2/\kappa_T$.

8.14 *Thermodynamic Identity, II.* Show that $C_P/C_V = \kappa_T/\kappa_S$.

8.15 *Extensive and Intensive Variables.* Divide the following fluid variables and functions of fluid variables, P, V, T, E, E^2, S, PV, H, T/H, κ_T, C_V, n, $c_V = C_V/n$ and $(\partial V/\partial T)_P$, into (1) extensive variables, (2) intensive variables, and (3) variables that are neither extensive nor intensive. State variables such as E^{-1} and $H \cdot S$ are neither extensive nor intensive.

8.16 *Joule Coefficient.* Start with the fundamental constraint $dE = TdS - PdV$, form each term into a derivative with respect to V with T held constant, and use reciprocity and one of the Maxwell relations to derive an

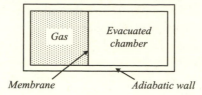

FIGURE 8.6 Expansion chamber. When the membrane is punctured, the gas expands into the evacuated chamber without doing work or absorbing heat. Used to measure the Joule coefficient $(\partial T/\partial V)_E$. (See Problem 8.16.)

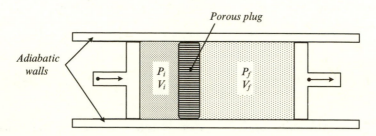

FIGURE 8.7 A system of gas to the left of the porous plug with volume V_i is maintained at pressure P_i by the left-hand piston. After it has passed through the plug, the gas is maintained at a lower pressure, $P_f < P_i$, with larger molar volume, $V_f/n_f > V_i/n_i$, by the right-hand piston. For this reason the total work performed on the gas is $P_iV_i - P_fV_f$. Since the channel walls are adiabatic, the gas exchanges no heat with its environment so that $Q = 0$. According to the first law of thermodynamics, the enthalpy of the gas is conserved, i.e., $E_i + P_iV_i = E_f + P_fV_f$, as it passes through the plug. (Used in Problem 8.17.)

expression for the so-called *Joule coefficient* $(\partial T/\partial V)_E = (P/C_V)(1 - T\alpha_V)$ where $\alpha_V \equiv P^{-1}(\partial P/\partial T)_V$.

The importance of this relation lies in its application to gases. The Joule coefficient can be measured in a chamber that allows a gas to expand isoenergetically into a vacuum. The coefficients C_V and α_V are directly related to the system's equations of state. See Figure 8.6.

8.17 *Joule-Thomson Coefficient.* Start with the fundamental constraint $dH = TdS + VdP$, form each term into a derivative with respect to P with

T held constant, and use one of the Maxwell relations to derive an expression for the so-called *Joule-Thomson coefficient* $(\partial T/\partial P)_H = (V/C_P)(T\alpha_p - 1)$ where $C_P = (\partial H/\partial T)_p$ and the isobaric expansivity $\alpha_p \equiv V^{-1}(\partial V/\partial T)_p$.

The importance of this relation lies in its application to gases. The Joule-Thomson coefficient is measured by forcing a gas through a channel partially stopped with a porous plug, as illustrated in Figure 8.7. The gas conserves its enthalpy as it moves through the plug. The coefficient C_P is readily measured, and the isobaric expansivity α_p is directly related to the system's equations of state. **(S)**

8.18 *Incompressible Solid*. Suppose a block of an incompressible solid (that is, with constant V) is subjected to an adiabatic reversible increase of pressure from P_i to P_f. Show that the ratio of its final temperature, T_f, to its initial temperature, T_i, is given by

$$\frac{T_f}{T_i} = \exp\left[\frac{V\alpha_p}{C_P}(P_f - P_i)\right].$$

NINE

Simple Fluid Systems

···

9.1 The Ideal Gas

···

The ideal gas law has been important for so long that it has achieved the status of an icon. For two centuries $PV = nRT$ was the only known equation of state. It universally and accurately describes low-density gases; no adjustable parameters are required to characterize individual gases in the low-density regime. As a result the pressure, P, of a constant-volume, low-density gas thermometer is a good indicator of thermodynamic temperature. Furthermore, its simplicity makes the ideal gas a favorite among theorists, teachers, and textbook writers. Even so the ideal gas is only one model among many.

The ideal gas is defined by two empirically descriptive rules: *Boyle's law* and *Joule's law*. Robert Boyle (1627–1691) found that *if the temperature of a gas is constant, the product PV is also constant*. In other words,

$$PV = f(T), \tag{9.1}$$

where $f(T)$ is an as yet undetermined function of temperature, T. Joule's law, that *the energy of a gas depends only on its temperature,*

$$E = E(T), \tag{9.2}$$

summarizes experiments performed much later, by Gay-Lussac in 1807 and more accurately by James Joule in 1845. In these experiments gases expanded freely and adiabatically into a vacuum. As the gas neither exchanged heat nor performed work on the environment, its expansion was an irreversible, energy-conserving process. Gay-Lussac and Joule found that the temperature of a gas changed very little during free expansion even as its pressure and volume changed significantly.

Of course, Boyle's and Joule's laws refer to empirical temperature. Only empirical temperatures are directly measurable, and Boyle, Gay-Lussac, and Joule (in 1845) knew nothing of thermodynamic temperatures. However, all useful temperatures are monotonic functions of each other.

Conditions (9.1)–(9.2) and the fundamental constraint for fluids, $dE = TdS - PdV$, are sufficient to generate the $P - V - T$ equation of state of an ideal gas. Note that Equations (9.1) and (9.2) link the variables P, V, T, and E. These variables are, for all kinds of fluid systems, either those measured or those related by very plausible assumptions. For this reason, the method we adopt here will be useful throughout Chapter 9. In particular, the cross-differentiation

$$\frac{\partial}{\partial V}\left(\frac{1}{T}\right)_E = \frac{\partial}{\partial E}\left(\frac{P}{T}\right)_V \tag{9.3}$$

arising from the entropy form of the fundamental constraint $dS = (1/T)dE + (P/T)dV$ provides the most directly useful relation imposed by thermodynamics.

Since, in this case, $E = E(T)$, T is constant when E is constant. Therefore, the left-hand side of Equation (9.3) vanishes. Using

Boyle's law, $P = f(T)/V$, to eliminate P from the right-hand side reduces (9.3) to

$$\frac{\partial}{\partial E}\left[\frac{f(T)}{T}\right]_V = 0, \tag{9.4}$$

which is equivalent to the total derivative

$$\frac{d}{dE}\left[\frac{f(T)}{T}\right] = 0. \tag{9.5}$$

According to the chain rule, (9.5) is equivalent to

$$\frac{d}{dT}\left[\frac{f(T)}{T}\right]\left(\frac{dT}{dE}\right) = 0, \tag{9.6}$$

which, ruling out $dT/dE = 0$, identifies $f(T)/T$ as a constant. In this way we find that PV/T is a constant independent of any of the thermodynamic variables of an ideal gas. Because V is an extensive variable and P and T are intensive variables, the constant PV/T must be proportional to an extensive quantity. If we choose $PV/T \propto n$, where n is the number of moles of gas in the system, then

$$PV = nRT, \tag{9.7}$$

where the proportionality constant $R = 8.31$ J/(K mole) is called the *gas constant*.

While $PV = nRT$ applies to all ideal gases, the energy equation of state $E = E(T)$ realizes different forms for different kinds of ideal gases. The simplest realization of $E = E(T)$ is

$$E = C_V T \tag{9.8}$$

where C_V is a state-independent heat capacity at constant volume. In this case the molar specific heat $c_V = C_V/n$ assumes dramatically different values for different types of ideal gases. For instance, $c_V = 3R/2$ for monatomic gases such as helium and argon, and $c_V = 5R/2$ for diatomic gases such as nitrogen, oxygen, and hydrogen. In general, $c_V = fR/2$, where f is the number of degrees of freedom (e.g., 3,5,6,...) of the particles composing the gas. The value of f and consequently the value of c_v shift abruptly at certain critical temperatures.

The two equations of state, $PV = nRT$ and $E = C_V T$, allow us to solve the identities (8.5a), $T = (\partial E / \partial S)_V$, and (8.5b), $P = -(\partial E / \partial V)_S$, for the energy-characterizing function $E = E(S, V)$. These equations become

$$\left(\frac{\partial E}{\partial S}\right)_V = \frac{E}{C_V} \tag{9.9}$$

and

$$\left(\frac{\partial E}{\partial V}\right)_S = \frac{-nRE}{VC_V}. \tag{9.10}$$

They integrate to

$$\ln E = \frac{S}{C_V} + f(V) \tag{9.11}$$

and

$$\ln E = \frac{-nR}{C_V} \ln V + g(S), \tag{9.12}$$

where $f(V)$ and $g(S)$ are functions whose consistency requires, apart from additive constants, that $f(V) = -(nR/C_V) \ln V$ and $g(S) = S/C_V$. Therefore,

$$\frac{E(S,V)}{E_o} = \frac{\exp\left\{\dfrac{S - S_o}{C_V}\right\}}{\left(\dfrac{V}{V_o}\right)^{nR/C_V}}, \tag{9.13}$$

where $E_o = E(S_o, V_o)$ or, equivalently,

$$S(E,V) = S_o + C_V \ln\left(\frac{E}{E_o}\right) + nR \ln\left(\frac{V}{V_o}\right), \tag{9.14}$$

where $S_o = S(E_o, V_o)$. Either of these functions, (9.13) or (9.14), completely characterizes an ideal gas with constant heat capacity C_V because both equations of state, (9.7) and (9.8), follow from appropriate derivatives.

Equation (9.14) immediately yields a very useful relation among the state variables of an ideal gas that has been reversibly and adiabatically transformed. Because, in this case, $dQ_{rev} = TdS = 0$, the entropy remains invariable, and $S(E,V) = S(E_o, V_o)$ in (9.14). The latter reduces (9.14) to $(E/E_o)^{C_V}(V/V_o)^{nR} = 1$, that is, to

$$E^{C_V}V^{nR} = \text{constant.} \tag{9.15}$$

The ideal gas equations of state, $E = C_V T$ and $PV = nRT$, allow us to eliminate the internal energy E in favor of the pressure P so that (9.15) further transforms to

$$PV^\gamma = \text{constant,} \tag{9.16}$$

where γ is defined by the ratio of heat capacities

$$\gamma \equiv \frac{C_P}{C_V}, \tag{9.17}$$

which, in turn, can be shown (see Problem 9.2) to be related to nR and C_V by $\gamma = 1 + nR/C_V$. (See Problems 9.1–9.9.)

9.2 Room-Temperature Elastic Solid

The distinguishing feature of a solid, apart from its rigidity, is its relative incompressibility. Most materials that are solid phase at standard temperature (273 K) and pressure (1 atm) change their volume very little as their temperature and pressure is diminished to zero. Therefore, in this regime, we expect the equation of state of a solid to assume a form $V = V(T,P)$ that we can accurately describe as a linear function of T and P with relatively small coefficients. Expanding $V = V(T,P)$ around $P = 0$ and $T = 0$ produces

$$V(T,P) \approx V_o + \left(\frac{\partial V}{\partial P}\right)_{T,o} P + \left(\frac{\partial V}{\partial T}\right)_{P,o} T, \tag{9.18}$$

where $V_o = V(0,0)$ and the partial derivatives $(\partial V/\partial P)_{T,o}$ and $(\partial V/\partial T)_{P,o}$ are evaluated at $P = 0$, $T = 0$. Equation (9.18) is equivalent to

$$V = V_o(1 + \alpha_{Po}T - \kappa_{To}P), \tag{9.19}$$

where the characterizing coefficients, $\alpha_{Po} \equiv V_o^{-1}(\partial V/\partial T)_{P,o}$ and $\kappa_{To} \equiv -V_o^{-1}(\partial V/\partial P)_{T,o}$, are evaluated at $P = 0$ and $T = 0$. Equivalently,

$$P = \frac{\alpha_{Po}}{\kappa_{To}}T + \frac{1}{\kappa_{To}}\left(1 - \frac{V}{V_o}\right). \tag{9.20}$$

This model incorporates, at least roughly, the behavior we expect of any solid material near room temperature and atmospheric pressure.

A complete thermodynamic description also requires information about how the internal energy, E, depends upon state variables. Most room-temperature solids have a state-independent heat capacity at constant volume C_V whose value is given by the law of Dulong and Petit, $C_V = 3nR$. Then $(\partial E/\partial T)_V = C_V$ integrates to

$$E = C_V T + f(V), \tag{9.21}$$

where the function $f(V)$, as we shall see, is determined by the requirement that the two equations of state, (9.20) and (9.21), are consistent with each other and the fundamental constraint. Again we use the entropy form of the fundamental constraint $dS = (1/T)$ $dE + (P/T)dV$ and exploit cross-differentiation

$$\frac{\partial}{\partial V}\left(\frac{1}{T}\right)_E = \frac{\partial}{\partial E}\left(\frac{P}{T}\right)_V. \tag{9.22}$$

Using the equation of state (9.20) to eliminate P from (9.22) produces

$$\frac{\partial}{\partial V}\left(\frac{1}{T}\right)_E = \frac{\partial}{\partial E}\left[\frac{\alpha_{Po}}{\kappa_{To}} + \frac{1}{\kappa_{To}T}\left(1 - \frac{V}{V_o}\right)\right]_V. \tag{9.23}$$

While the independent variables associated with the derivatives are E and V, the quantities on which they operate are functions of T and V. One straightforward way to proceed is to solve the energy equation (9.21) for $T = T(E,V)$, replace T in (9.23) with $T(E,V)$, and complete the derivatives. Equivalently, and more easily, we can use the chain rule to perform the derivatives implicitly so that (9.23) becomes

$$\frac{1}{T^2}\left(\frac{\partial T}{\partial V}\right)_E = \frac{1}{T^2 \kappa_{To}}\left(1 - \frac{V}{V_o}\right)\left(\frac{\partial T}{\partial E}\right)_V. \tag{9.24}$$

The derivatives $(\partial T/\partial V)_E$ on the left side of (9.24) and $(\partial T/\partial E)_V$ on the right side of (9.24) are linked by the reciprocity rule $(\partial T/\partial V)_E$ $(\partial V/\partial E)_T (\partial E/\partial T)_V = -1$, applied to the variables T, V, and E. This reciprocity rule reduces (9.24) to

$$\left(\frac{\partial E}{\partial V}\right)_T = \frac{-1}{\kappa_{To}}\left(1 - \frac{V}{V_o}\right). \tag{9.25}$$

Integrating this result produces

$$E = \frac{(V^2 - 2VV_o)}{2\kappa_{To}V_o} + g(T). \tag{9.26}$$

This equation and $E = C_V T + f(V)$ are consistent only if, apart from an arbitrary constant,

$$E = C_V T + \frac{(V - V_o)^2}{2\kappa_{To}V_o}. \tag{9.27}$$

Apparently, elastic energy is stored in the solid as its volume departs from its reference value, V_o. Equations of state (9.20) and (9.27) completely describe this model of a room-temperature solid. (See Problem 9.10.)

9.3 Cavity Radiation

Material bodies continually absorb and re-emit electromagnetic radiation. If you have stood in front of a blazing fire on a cold night, you have felt your skin absorbing electromagnetic radiation. When emitting and absorbing bodies are not in mutual equilibrium, such emission and absorption is an irreversible process. However, when emitting and absorbing bodies are in mutual equilibrium, the radiation can be characterized as a thermodynamic fluid with temperature, energy, volume, entropy, and pressure.

A simple way to imagine such *equilibrium* or *cavity radiation* is to visualize, as in Figure 9.1, an evacuated cavity of volume V surrounded by a material with temperature T. The cavity radiation and the surrounding material achieve mutual equilibrium as radiation is continually exchanged across the cavity. When cavity radiation reversibly absorbs heat dQ_{rev} from the walls, its entropy increases by $dS = dQ_{rev}/T$. When its walls expand, cavity radiation performs PdV reversible work. In every way cavity radiation seems well described by the fluid variables P, V, T, E, and S, which, in turn, are related to each other by the fundamental constraint $dE = TdS - PdV$.

Equations of state follow directly from the laws of electromagnetism and thermodynamics. According to Maxwell's theory of

FIGURE 9.1　A cavity filled with equilibrium radiation.

electromagnetic radiation a beam of unidirectional light exerts a pressure P equal to its energy density E/V when absorbed. But when the radiation is isotropic, as in a cavity,

$$P = \frac{1}{3}\frac{E}{V}.$$

(9.28)

As we shall see, the energy density E/V is a function of temperature T alone, that is,

$$\frac{E}{V} = e(T),$$

(9.29)

where $e(T)$ is an as yet undetermined universal function of temperature.

Any dependence other than (9.29) would violate the second law of thermodynamics. If, for instance, $E/V = e(T,X)$ where X was, say, the volume or parameterized the shape of the cavity or the condition of its walls, then we could arrange for two cavities, A and B, with the same temperatures, T, and different values of X, such that $e(T,X_A) > e(T,X_B)$, to exchange radiation through a small connecting tube. Since the flux of radiant energy is proportional to local energy density, net energy would be transferred from cavity A to cavity B. This heat flux would increase the temperature of cavity B and decrease the temperature of A and so violate the Clausius statement of the second law.

The cavity walls are both a source and a sink of cavity radiation. If the cavity volume expands isothermally, its walls emit the cavity radiation needed to keep E/V constant. If the cavity volume contracts isothermally, the walls absorb cavity radiation. Such behavior distinguishes a cavity filled with radiation from a cavity filled with gas. Gas and cavity radiation are both fluids. But gas is a conserved quantity; cavity radiation is not.

There are a number of ways to solve the two equations of state, $P = E/3V$ and $E/V = e(T)$, and the fundamental constraint $dE = TdS - PdV$ for the function $e(T)$. We could again use cross-differentiation resulting from the entropy form of the fundamental constraint, but the result also falls out quickly if we simplify the equations of state by shifting independent variables from (E, V) to (e, V), where $e = E/V$. Then $P = E/3V$ becomes

$$P = \frac{e}{3} \tag{9.30}$$

and the fundamental constraint

$$d(eV) = TdS - PdV. \tag{9.31}$$

Using Equation (9.30) to eliminate P from the constraint (9.31) produces

$$dS = \frac{4e}{3T}dV + \frac{V}{T}de. \tag{9.32}$$

Therefore, $(\partial S/\partial V)_e = 4e/3T$ and $(\partial S/\partial e)_V = V/T$. Cross differentiation produces

$$\frac{\partial}{\partial e}\left(\frac{4e}{3T}\right)_V = \frac{\partial}{\partial V}\left(\frac{V}{T}\right)_e. \tag{9.33}$$

Since T is constant when e is constant, this equation reduces to

$$\left(\frac{\partial e}{\partial T}\right)_V = \frac{4e}{T}, \tag{9.34}$$

which integrates to

$$e(T) = aT^4, \tag{9.35}$$

that is, to

$$E/V = aT^4. \tag{9.36}$$

Integration of Equation (9.34) allows for a to be an arbitrary function of V, but we also know that E/V is a function of T alone. The universal constant a is called the *radiation constant*.

Cavity radiation is ubiquitous and universal. Its equations of state contain no adjustable parameters. All objects in thermal equilibrium produce such radiation whether they contain extended cavities or not. All objects in thermal equilibrium necessarily absorb the same frequencies they radiate. For this reason cavity radiation is also called *black-body radiation*. The sun, for instance, emits an observable flux of cavity radiation equal to one-fourth the speed of light in a vacuum, $c/4$, times aT^4, that is, $\sigma_B T^4$ where the *Stefan-Boltzmann constant* $\sigma_B (\equiv ca/4)$. When further analyzed, cavity radiation supplies important clues as to how atoms and molecules absorb and emit electromagnetic radiation. (See Problems 9.11–9.16.)

Chapter 9 Problems

9.1 *Ideal Gas, I*. Show that, for an ideal gas with constant C_V, the adiabatic compressibility $\kappa_S = P^{-1}(1 + nR/C_V)^{-1}$.

9.2 *Ideal Gas, II*. Show that, for an ideal gas with $C_V = fnR/2$, where f is the number of degrees of freedom of the molecules composing the gas,
(a) $C_P - C_V = nR$;
(b) $C_P/C_V = 1 + 2/f$, where $\gamma \equiv C_P/C_V$ and $\gamma = 1 + 2/f$.

9.3 *Adiabatic Transformation*. A reversible adiabatic transformation is one for which $dQ_{rev} = 0$. Starting from the first law of thermodynamics for reversible, adiabatic transformations of an ideal gas show that
(a) During a reversible adiabatic transformation of an ideal gas the following are constant: $TV^{\gamma-1}$, PV^γ, and $P^{1-\gamma}T^\gamma$. Here $\gamma \equiv C_P/C_V$. (S)
(b) The expression for the entropy of an ideal gas $S(E,V)$ reduces to a constant when the results of part (a) are substituted into the expression $S(E,V)$.

9.4 *Polytropes of an Ideal Gas*. If we require the ratio dQ/dT to be a constant C under all conditions, the first law of thermodynamics applied to a fluid $dE = dQ - PdV$ reduces to the constraint $dE = CdT - PdV$. Solutions of this constraint are called *polytropes*, and when $PV = nRT$ and $E = C_V T$

they are called *polytropes of an ideal gas*. Polytropes are used extensively in modeling stellar structures.

(a) Exploiting what we know about ideal gases from Problems 9.2 and 9.3, derive a quantity that is constant during the reversible polytropic expansion (or contraction) of an ideal gas.

(b) Show that when $dQ = 0$ (that is, when $C = 0$), this constant quantity reduces to the quantity held constant during an adiabatic expansion or contraction.

(c) Show that the constant found in (a) reduces to a reasonable result when $C = C_V$.

(d) Show that the constant found in (a) reduces to a reasonable result when $C = C_p$.

9.5 *Adiabatic Expansion*. Three moles of O_2 gas at an initial temperature of 400°C and a pressure of 25 atm are expanded adiabatically until the final temperature is 100°C. Find the final pressure. Assume that O_2 is an ideal gas with $\gamma = 7/5$.

9.6 *Adiabatic Compression in a Diesel Engine*. A diesel engine requires no spark plug. Rather, its fuel ignites spontaneously when sprayed into the highly compressed air of an engine piston. Suppose that initially the air within a diesel engine piston is at 1 atm and that the engine piston adiabatically compresses the air by a factor of 15. This factor, by which the piston volume is decreased, is called the *compression ratio*. The ratio of specific heats for air $C_p/C_V [\equiv \gamma] = 7/5$. Find the pressure of the completely compressed air.

9.7 *Carnot Cycle with Ideal Gas as Working Fluid*. Suppose n moles of ideal gas for which $E = C_V T$ is the working fluid of a Carnot cycle operating between reservoirs with temperatures T_C and $T_H > T_C$, as illustrated in Figure 9.2. The subscripts 1, 2, 3, and 4 denote the states at which adiabats and isotherms intersect. Thus, $T_H = T_1 = T_2$, $T_C = T_3 = T_4$, $S_2 = S_3$, and $S_4 = S_1$. Also $\gamma \equiv C_p/C_V$. Determine the following in terms of n, T_H, T_C, V_1, V_2, V_3, V_4, and C_V.

(a) The work $W_{1 \to 2}$ done by the ideal gas during the isothermal process $1 \to 2$

(b) The heat Q_H absorbed by the ideal gas during the isothermal process $1 \to 2$

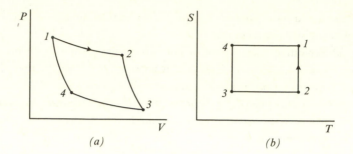

(a)   (b)

FIGURE 9.2   State variable diagrams for Carnot cycle. (Used in Problems 9.7 and 9.14.)

(c) The work $W_{2\to3}$ done by the ideal gas during the adiabatic process $2 \to 3$

(d) The total work W done by the ideal gas in one cycle

(e) The efficiency W/Q_H of the cycle implied by the answers to parts (b) and (d). Show that this efficiency is consistent with the efficiency, $1 - T_C/T_H$, of a Carnot cycle.

[Hint: You will have to use the equations of state for an ideal gas and the results of Problem 9.3a.]

9.8 *Joule Expansion*. An ideal gas for which $E = C_V T$ is confined to the left side of a chamber surrounded by an adiabatic wall, as shown in Figure 8.6. The right-hand side is evacuated. The partition is punctured, and the fluid quickly and irreversibly occupies the whole chamber. The initial state volume and pressure of the n moles of this gas are V_i and P_i. The final volume is V_f. Express the following in terms of V_i, V_f, P_i, and n.

(a) The change in energy ΔE of the ideal gas

(b) The change in entropy ΔS

(c) The final pressure P_f

(d) The final temperature T_f

9.9 *Air Standard Otto Cycle*. The cylinders of a gasoline engine contain rapidly burning fuel, reject waste gases, and draw in replacement fuel during different parts of a highly irreversible, nonequilibrium cycle. Even so, important features of the gas engine cycle can be modeled and discovered by analyzing the reversible *air standard Otto cycle* illustrated in Figure 9.3. The working fluid is an ideal diatomic gas for which $\gamma [\equiv C_P/C_V] = 7/5$

and $E = C_V T$. The cycle consists of four strokes: (1) *compression* $1 \rightarrow 2$ during which the piston rapidly, and thus adiabatically, compresses the fluid from its largest, V_L, to its smallest volume, V_S; (2) *explosion* $2 \rightarrow 3$ during which heat is rapidly and isochorically deposited in the fluid raising its temperature from T_2 to $T_3 > T_2$ while $V = V_S$; (3) *power* $3 \rightarrow 4$ during which the piston rapidly and adiabatically expands from V_S to V_L, delivering work to the environment; and (4) *exhaust* $4 \rightarrow 1$ during which the fluid isochorically rejects heat to a continuous series of reservoirs with temperatures between T_4 and $T_1 < T_4$ while $V = V_L$. During the adiabatic strokes PV^γ and $TV^{\gamma-1}$ remain constant, while during the isochoric strokes V remains constant. The processes $1 \rightarrow 5 \rightarrow 1$ during which, in a more complete cycle, waste is rejected and fuel replenished are not part of the air standard Otto cycle. Express answers in terms of the temperatures T_1, T_2, T_3, and T_4, the heat capacity at constant volume C_V, and the ratio of specific heats $\gamma \equiv C_P/C_V$. Find the following:

(a) The adiabatic work done on the fluid during the compression stroke

(b) The heat isochorically absorbed by the fluid during the explosion stroke

(c) The adiabatic work performed by the fluid on its environment during the power stroke

(d) The heat rejected by the fluid during the exhaust stroke

(e) The relative ordering of temperatures T_1, T_2, T_3, and T_4
[Hint: see Figure 9.3.]

(f) The cycle efficiency—that is, the ratio of work delivered to heat absorbed—in terms of the compression ratio $r \equiv V_L/V_S$
[Hint: Use the relationship between the heat capacity at constant volume C_V, nR, and γ for an ideal gas.]

(g) The numerical value of the efficiency for a compression ratio $r = 10$

9.10 *Room-Temperature Solid.* The energy-characterizing function for a room-temperature solid is

$$E(S,V) = \frac{(V - V_o)^2}{2\kappa_{To} V_o} + E_o \exp\left\{\frac{S}{C_V} - \frac{\alpha_{Po} V}{C_V \kappa_{To}}\right\},$$

where E_o is an arbitrary integration constant with energy units.

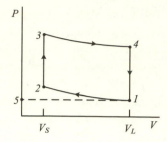

FIGURE 9.3 Reversible air standard Otto cycle. The solid line $1 \to 2 \to 3 \to 4 \to 1$ represents the air standard Otto cycle. The dashed line $1 \to 5 \to 1$ represents exhaust and intake strokes. (Used in Problem 9.9.)

(a) Show that the two equations of state for a room-temperature solid, $P = \alpha_{P_o}T/\kappa_{T_o} + (1 - V/V_o)/\kappa_{T_o}$ and $E = C_V T + (V - V_o)^2/(2\kappa_{T_o}V_o)$, can be generated from $E(S,V)$ and appropriate derivatives of $E(S,V)$. **(S)**

(b) [More difficult] Integrate these two equations of state of the room-temperature solid and, in this way, generate its energy-characterizing function.

9.11 *Cavity Radiation, I.* Show that the energy equation of state for cavity radiation, $E/V = aT^4$, where a is a constant, follows directly from the pressure equation of state, $P = E/3V$, the fundamental constraint $dE = TdS - PdV$, and the condition $(\partial/\partial V)(E/V)_T = 0$.

9.12 *Cavity Radiation, II*

(a) Show that the characterizing function $E(S,V)$ for cavity radiation is given by $E(S,V) = (3S/4)^{4/3}/(aV)^{1/3}$. **(S)**

(b) Show that its two equations of state, (9.28) and (9.36), follow from appropriate derivatives of this function.

9.13 *Adiabatic Cavity Radiation.* Show that when a region filled with cavity radiation quasistatically and adiabatically expands or contracts, the quantities $TV^{1/3}$ and $PV^{4/3}$ remain constant.

9.14 *Carnot Cycle with Cavity Radiation as Working Fluid.* Cavity radiation, described by equations of state $P = E/3V$ and $E/V = aT^4$, is the working fluid of a Carnot cycle operating between reservoirs with temperatures T_C and $T_H > T_C$, as illustrated in Figure 9.2. The subscripts 1, 2, 3, and

4 denote the states at which adiabats and isotherms intersect. Thus $T_H = T_1 = T_2$, $T_C = T_3 = T_4$, $S_2 = S_3$, and $S_4 = S_1$. Determine the following in terms of T_H, T_C, V_1, V_2, V_3, and V_4.

(a) The work $W_{1 \to 2}$ done by the cavity radiation during the isothermal expansion $1 \to 2$

(b) The heat Q_H absorbed by the cavity radiation during the isothermal expansion $1 \to 2$

(c) The work $W_{2 \to 3}$ done by the cavity radiation during the adiabatic process $2 \to 3$

(d) The total work W done by the cavity radiation during one cycle

(e) The efficiency W/Q_H of the cycle implied by answers to parts (b) and (d). Show that this efficiency is consistent with the efficiency, $1 - T_C/T_H$, of a Carnot cycle.

[Hint: You will have to use the equations of state for cavity radiation and the other results of Problem 9.12.]

9.15 *Solar Pressures.* The sun may be considered a sphere of ideal gas composed of neutral molecules, electrons, and ions with a superimposed sphere of equilibrium cavity radiation. The gas density and temperature vary from the center to the surface of the sphere. Find the ideal gas pressure and the pressure of the cavity radiation at the following places within the sun; express these pressures in atmospheres; and form the ratio of ideal gas pressure to cavity radiation pressure.

(a) The solar center at which the mole number density of solar gas (neutral molecules, electrons, and ions) is 8.3×10^7 moles/m^3 and the temperature 1.5×10^7 K

(b) The photosphere—that is, the visible solar surface—at which the mole number of solar gas is 0.17 moles/m^3 and the temperature is 5800 K

9.16 *Solar Constant.* The solar flux originates at the solar photosphere, spreads uniformly in all directions, and eventually strikes the earth. Calculate the *solar constant*, that is, the magnitude of the solar flux at the earth in W/m^2. Recall that the flux of cavity radiation is $c/4$ times the cavity radiation energy density E/V. Use the data from Problem 9.15 and the facts that the solar radius $R_S = 6.96 \times 10^8$ m and the mean sun-earth distance is $R_{SE} = 1.50 \times 10^{11}$ m.

TEN

Nonfluid Systems

··

10.1 Nonfluid Variables

··

The analytical methods presented thus far apply generally to any kind of system and not only to those described with fluid variables. Nonfluid systems retain temperature, T, internal energy, E, and entropy, S, as variables but differ in the way reversible work is done. Every fundamental constraint assumes the form

$$dE = TdS + \sum_i F_i dX_i, \tag{10.1}$$

where $\sum_i F_i dX_i$ is the total reversible work done on the system and $F_i dX_i$ is one among its several possible contributions. In order to fully describe each system we need to know each *generalized force*, F_i; its *associated displacement*, X_i; and a set of equations of state, $T = T(S, X_1, X_2, ...)$ and $F_i = F_i(S, X_1, X_2, ...)$, for each i or, alternatively, a characterizing energy as a function of its proper variables, say, $E(S, X_1, X_2, ...)$. When reversible work is done in only one way, the internal energy is a function of two independent variables—the entropy, S. and a displacement, X. But multivariate systems are quite possible.

10.2 The Theoretician's Rubber Band

The theoretician's rubber band is a one-dimensional system of length L that exists only in tension. We represent this tension with an inherently positive variable $F > 0$. The differential reversible work done on the system dW_{rev} when its length is increased differentially by dL is

$$dW_{rev} = FdL. \qquad (10.2)$$

Therefore, its fundamental constraint is

$$dE = TdS + FdL; \qquad (10.3)$$

its associated equations of state assume the forms $T = (\partial E/\partial S)_L$ and $F = (\partial E/\partial L)_S$; and the associated Maxwell relation is $(\partial T/\partial L)_S = (\partial F/\partial S)_L$.

Interestingly, the thermal expansivity of rubber in constant tension is negative, that is, $(\partial L/\partial T)_F < 0$, as can be demonstrated by hanging a weight from a length of rubber band, heating the rubber band with a match, and observing the weight rise. In other words, assuming that the rubber band behaves like a Hooke's Law spring with $F \propto L$, its spring constant is an increasing function of temperature. The simplest equation of state that incorporates this behavior is

$$F = \alpha TL, \qquad (10.4)$$

where the constant $\alpha > 0$. This equation of state limits, but does not determine, the form assumed by the energy equation of state.

In order to relate the internal energy E to the variables appearing in Equation (10.4)—F, T, and L—we isolate terms containing these variables on the right-hand side of the fundamental constraint, so that $dS = (1/T)dE - (F/T)dL$, and pick out the cross-differentiation

$$\frac{\partial}{\partial L}\left(\frac{1}{T}\right)_E = -\frac{\partial}{\partial E}\left(\frac{F}{T}\right)_L. \tag{10.5}$$

Given the tensile equation of state $F = \alpha TL$, relation (10.5) implies $(\partial T/\partial L)_E = 0$. According to this result, temperature, T, is not a function of L. Consequently, T must be a function of internal energy, E, alone, or, conversely,

$$E = E(T). \tag{10.6}$$

This result is unexpected. For Equation (10.6) claims that stretching a rubber band slowly, that is, isothermally, does not increase its internal energy. Apparently, isothermal stretching forces heat out of the rubber band at the same rate that work is done on it.

Other interesting facts about rubber bands follow readily from the entropy form of the fundamental constraint (10.3), $dS = (1/T) dE - (F/T)dL$, and the tensile equation of state (10.4), $F = \alpha TL$. For instance,

$$\left(\frac{\partial S}{\partial L}\right)_E = \frac{-F}{T} < 0. \tag{10.7}$$

Apparently, the entropy of a rubber band decreases as its length is quasistatically increased when its energy, that is, temperature, is held constant. Another interesting fact follows from the chain rule

$$\left(\frac{\partial T}{\partial L}\right)_S = \left(\frac{\partial T}{\partial E}\right)_S\left(\frac{\partial E}{\partial L}\right)_S. \tag{10.8}$$

From the fundamental constraint $dE = TdS + FdL$, we know that $(\partial E/\partial L)_S = F$. Given $E = E(T)$ and the reciprocal rule $(\partial T/\partial E)_S = 1/(\partial E/\partial T)_S$, Equation (10.8) becomes

$$\left(\frac{\partial T}{\partial L}\right)_S = \frac{F}{(dE/dT)}. \tag{10.9}$$

Since we have specified that $F > 0$, and, as we will find in Chapter 12, $dE/dT > 0$ is required for intrinsic stability,

$$\left(\frac{\partial T}{\partial L}\right)_S > 0. \tag{10.10}$$

Apparently, stretching a rubber band quasistatically and adiabatically increases its temperature. In practice, adiabatic stretching of a rubber band must be done quickly in order to prevent heat from entering or leaving the band. (See Problem 10.1.)

10.3 Paramagnetism

Materials composed of atoms having a magnetic dipole moment respond in various ways to an applied magnetic field. When only a small fraction of the atomic dipoles align themselves with the applied magnetic field (the usual case), the material is called *paramagnetic*. On the other hand, a relatively large fraction of the atomic dipoles within small regions, or *domains*, of *ferromagnetic* materials spontaneously align with each other even in the absence of an applied field. In the presence of an applied field the ferromagnetic domains themselves become aligned with one another and with the applied field. Here we limit our discussion to paramagnetism.

Figure 10.1 sketches an experimental apparatus designed for studying the response of a paramagnetic material. A current-carrying solenoid creates a relatively uniform magnetic field, B_o, in the interior of the solenoid into which a sample of paramagnetic material has been inserted. This field exerts a torque on the atomic dipoles and causes them to align, in varying degrees, with B_o. These dipoles, in turn, create a magnetic field, B_m, that adds to the applied field, B_o, and creates a net field

$$B = B_o + B_m \tag{10.11}$$

within the material.

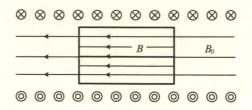

FIGURE 10.1   A sample of magnetic material inside a solenoid.

Paramagnetic materials are thermodynamic systems. They can exchange heat and work with their environment. The currents that create B_o do reversible work

$$dW_{rev} = \frac{B_o}{\mu_o} dB_m \qquad (10.12)$$

on the paramagnetic material, where here, μ_o is the so-called *permittivity of free space*. If only magnetic work can be done on the system, its fundamental constraint is

$$dE = TdS + \frac{B_o}{\mu_o} dB_m; \qquad (10.13)$$

the associated equations of state assume the forms $T = (\partial E/\partial S)_{B_m}$ and $B_o/\mu_o = (\partial E/\partial B_m)_S$; and cross-differentiation of these produces $\mu_o(\partial T/\partial B_m)_S = (\partial B_o/\partial S)_{B_m}$. One often finds reversible magnetic work expressed as $dW_{rev} = \mu_o H dM$, where $H [\equiv B_o/\mu_o]$ is the magnetic induction and $M [\equiv B_m/\mu_o]$ the magnetic moment per unit volume. The notations H and M emphasize the utterly different sources of these fields, but here I find it more suggestive to retain the symbols B_o and B_m because they remind us of the common nature of these fields.

In general, the response field B_m is an ever-increasing function, $B_m(B_o)$, of the applied field, B_o. Pierre Curie (1850–1906) experimentally discovered an equation of state for paramagnetic materials,

$$B_m = \mathscr{C} \frac{B_o}{T}, \qquad (10.14)$$

where the *Curie constant*, $\mathscr{C}$, characterizes the material. *Curie's law* (10.14) expresses the competing effects of the applied field, B_o, and the temperature, T; the former promotes while the latter inhibits magnetization. Later, Paul Langevin derived a more general equation of state for paramagnetic materials from a statistical model, but Curie's law remains valid for the regime of relatively small magnetizations and high temperatures. We confine our analysis to this regime.

Interestingly, the reversible work and equation of state expressions of the theoretician's rubber band and of a paramagnetic material are structurally identical. Compare Equation (10.2) with (10.12) and (10.3) with (10.13). For this reason deductions from the former can be transformed into deductions from the latter via the transformations $F \to B_o/\mu$ and $L \to B_m$. In particular, the energy of a paramagnetic material, like that of a rubber band, is a function of temperature alone, that is, $E = E(T)$.

The heat capacity of a paramagnetic material in a constant applied field, that is,

$$C_{B_o} = T \left(\frac{\partial S}{\partial T} \right)_{B_o},$$

(10.15)

enters into the expression

$$\left(\frac{\partial T}{\partial B_o} \right)_S = \frac{\mathscr{C} B_o}{\mu_o T C_{B_o}}$$

(10.16)

(derived below), which, in turn, encapsulates the idea behind the technique of *adiabatic demagnetization*. Since each quantity on the right-hand side of Equation (10.16) is positive, $(\partial T/B_o)_S > 0$. Therefore, the temperature, T, of the paramagnetic system decreases as the applied field, B_o, is quasistatically turned down. Scientists have used adiabatic demagnetization to achieve laboratory temperatures below 10^{-6} K.

Equation (10.16) and the inequality $(\partial T/B_o)_S > 0$ that it implies follow from the reciprocity rule applied to the variables T, B_o, and S:

$$\left(\frac{\partial T}{\partial B_o}\right)_S \left(\frac{\partial B_o}{\partial S}\right)_T \left(\frac{\partial S}{\partial T}\right)_{B_o} = -1. \qquad (10.17)$$

The reciprocal rule and Equation (10.15) transform (10.17) into

$$\left(\frac{\partial T}{\partial B_o}\right)_S = \frac{-T\left(\frac{\partial S}{\partial B_o}\right)_T}{C_{B_o}}. \qquad (10.18)$$

Furthermore, one can show that $(\partial S/\partial B_o)_T = \mu_o^{-1}(\partial B_m/\partial T)_{B_o}$ is one of the four cross-differentiations following from the magnetic system's fundamental constraint Eq. (10.13). This relation and Curie's law, (10.14), transform (10.18) into the desired result (10.16). (See Problems 10.2 and 10.3.)

10.4 Surfaces

Surface tension is a macroscopic manifestation of intermolecular attraction. Its effects are especially dramatic in liquids, where intermolecular attraction tends to minimize surface area. Small droplets assume the shape of a perfect sphere, since that shape minimizes the surface area of a given volume.

We conceptualize a surface as a distinct thermodynamic system with its own internal energy, E; temperature, T; and entropy, S. In place of a volume, a surface has an area, A. One way to define surface tension, σ, is to say that the reversible work done on a surface in changing its area by dA is

$$dW_{rev} = \sigma dA. \qquad (10.19)$$

Equivalently, surface tension is the force that a surface exerts normal to its edge per unit edge length in its local plane, as illustrated in

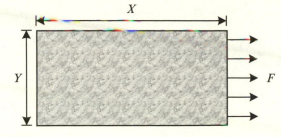

FIGURE 10.2 The surface tension σ is that force, F, exerted by or on a surface in its plane normal to its edge divided by the edge length, Y, that is, $\sigma = F/Y$. The work done on the surface in increasing its length by dX is FdX, that is, $(F/Y)YdX$ or σdA.

Figure 10.2. Given Equation (10.19), the fundamental constraint of a surface is $dE = TdS + \sigma dA$.

In determining the equations of state of any system we appeal either to observations of the system or to a reasonable theoretical model of the system or to both. Here I choose to start with a very simple theoretical model of a surface and deduce what relations follow from that model.

Figure 10.3a shows the cross section of a surface on a molecular scale. Imagine that each molecule strongly attracts and is strongly attracted to its nearest neighbors yet also strongly resists intermolecular penetration. In this way the molecules slide around each other and always maintain a given spacing. Therefore the average density of molecules within the bulk phase and on the surface remains constant. As the surface stretches, molecules rise up from the bulk to occupy new openings on the surface; the bulk volume shrinks and the surface area grows. As the surface contracts, molecules in the surface sink into the bulk phase below. Of course, this model describes only certain kinds of surfaces.

According to the model the net intermolecular force seen by one molecule is roughly described by the potential well shown in Figure 10.3b. Within the bulk phase, net forces on a molecule vanish. Equivalently, a molecule may move freely on the floor of the potential well. When drawn up into the surface—that is, to the po-

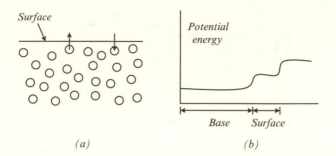

FIGURE 10.3 (a) Molecular-scale cross section of a surface. Molecules move into or out of the surface as the surface expands or contracts. (b) Associated energy level diagram.

tential shelf—a molecule overcomes net attractive forces pulling it back into the bulk. To escape the surface altogether a molecule must overcome the net attractive forces pulling it back into the surface. The potential shelf defining the surface is approximately halfway between the bulk phase potential floor and the field-free potential beyond the surface.

Thus, the energy of the surface is directly proportional to the number of molecules it contains. Since these are necessarily evenly distributed, surface energy is also proportional to surface area, that is,

$$E = f(T)A, \tag{10.20}$$

where the function $f(T)$ is not yet known. Furthermore, since each molecule interacts only with its nearest neighbors, the surface tension, σ, is independent of the surface area, A, that is,

$$\sigma = \sigma(T). \tag{10.21}$$

The fundamental constraint $dE = TdS + \sigma dA$, given (10.20), relates the two unknown functions $f(T)$ and $\sigma(T)$. Because conditions (10.20) and (10.21) are expressed in terms of σ, A, T, and E we again use the cross-multiplication arising from the entropy form, $dS = (1/T)dE - (\sigma/T)dA$, of the fundamental constraint, that is,

$$\frac{\partial}{\partial A}\left(\frac{1}{T}\right)_E = -\frac{\partial}{\partial E}\left(\frac{\sigma}{T}\right)_A. \tag{10.22}$$

Performing these derivatives implicitly produces

$$-\frac{1}{T^2}\left(\frac{\partial T}{\partial A}\right)_E = -\frac{d}{dT}\left[\frac{\sigma(T)}{T}\right]\left(\frac{\partial T}{\partial E}\right)_A. \tag{10.23}$$

The reciprocity relation applied to the variables T, A, and E and the reciprocal relation converts (10.23) into

$$\left(\frac{\partial E}{\partial A}\right)_T = -T^2 \frac{d}{dT}\left[\frac{\sigma(T)}{T}\right]. \tag{10.24}$$

The equation of state $E = f(T)A$, in turn, transforms this result into

$$f(T) = -T^2 \frac{d}{dT}\left[\frac{\sigma(T)}{T}\right], \tag{10.25}$$

that is, into

$$f(T) = \sigma(T) - T\frac{d\sigma(T)}{dT}. \tag{10.26}$$

Therefore, the energy equation of state (10.20) assumes the form

$$E = \left[\sigma(T) - T\frac{d\sigma(T)}{dT}\right]A \tag{10.27}$$

with the surface tension function $\sigma(T)$ remaining unspecified.

To specify the function $\sigma(T)$ requires new physics. For instance, we know that above a certain critical temperature, T_c, surface tension vanishes and neither liquids nor their surfaces exist. Therefore, this model should be valid only for $T \leq T_c$. Several empirically motivated equations of state with this property have been investigated. The so-called Eötvös and Ferguson energy equations of state are taken up in the problems. Here we introduce the simple assumption that $f(T) = cT$, that is,

$$E = cAT, \tag{10.28}$$

where c is a constant specific heat for constant surface area [$\equiv A^{-1}(\partial E/\partial T)_A$]. This choice leads, via integration of (10.27) with the initial condition $\sigma(T_C) = 0$, to

$$\sigma(T) = cT \ln\left(\frac{T_C}{T}\right) \quad \text{for } 0 \le T \le T_C \tag{10.29}$$

and $\sigma(T) = 0$ when $T > T_C$. (See Problems 10.4–10.7.)

10.5 Chemical Potential

The model of a surface just outlined associates changes in surface area with the absorption or rejection of its matter. If each mole added to or subtracted from a surface increases or decreases its area by a fixed amount, A_o, the number of moles in the surface, n, is related to its area, A, by

$$n = \frac{A}{A_o}, \tag{10.30}$$

where the constant A_o depends only upon the kind of molecule composing the surface. If we use Equation (10.30) to replace surface area A with mole number n in our description, the fundamental constraint of a surface becomes

$$dE = TdS + \mu dn, \tag{10.31}$$

where the intensive variable $\mu \equiv A_o\sigma$ is called the *chemical potential*—something of a misnomer that suggests, incorrectly, that chemical reactions are necessarily the agent of mole number change. Here the energy has become a function, $E(S,n)$, of the independent variables, S and n, so that $T = (\partial E/\partial S)_n$ and $\mu = (\partial E/\partial n)_S$.

Transforming the thermodynamic description of a surface in this way changes nothing except our point of view—but point of view is important. The idea that the internal energy can change not only through heat exchange and work, but also through the transfer of matter, expands the range of phenomena described by thermodynamics to include *open systems*. This change of view originates with the work of J. Willard Gibbs (1839–1903), who may have been the greatest mathematical physicist the United States has ever produced. Gibbs made important contributions to electromagnetic theory, vector analysis, and statistical mechanics as well as to thermodynamics.

Our model of a surface has allowed us to replace one of its independent variables with system mole number. But most important realizations of open systems are of composite or heterogeneous systems whose different parts are each open subsystems. Matter in each subsystem can be transformed into matter in other subsystems either by phase change (for instance, from liquid to vapor) or by chemical reaction.

To model a homogeneous, open fluid subsystem we simply add its mole number, n, to the list of its independent variables so that $E = E(S,V,n)$ and append the term μdn to its fundamental constraint so that

$$dE = TdS - PdV + \mu dn. \tag{10.32}$$

The term μdn describes how the addition $(dn > 0)$ or subtraction $(dn < 0)$ of moles changes the subsystem energy. Because the mole number n is now an independent variable, deductions from Equation (10.32) must be appropriately generalized. Among those taking the form of an equation of state are $T = (\partial E/\partial V)_{V,n}$, $-P = (\partial E/\partial V)_{S,n}$, and

$$\mu = \left(\frac{\partial E}{\partial n}\right)_{S,V}. \tag{10.33}$$

Equation (10.33) describes a relation among system variables, including their dependence on system mole number, n. For this reason we expect the function $\mu(S,V,n)$ to be closely related to the physics of a closed system for which n is constant.

Just how $\mu(S,V,n)$ might follow from the characterizing energy of a closed system $E(S,V)$ in which the mole number, n, appears only as a parameter depends upon the crucial distinction between extensive and intensive variables. When a fluid system simply changes its size while keeping its intensive variables, T, P, and μ, constant, each of its extensive variables, E, S, V, and n, changes by the same fraction. For instance, if n increases by 2% and the intensive variables remain constant, the extensive variables E, S, and V each increase by 2%. Likewise, if n changes by the differential amount dn while the intensive variables T, P, and μ are held constant, the ratio dn/n must equal each of the ratios dE/E, dS/S, and dV/V. Denoting this differential-sized ratio as α, we have $\alpha = dE/E$, $\alpha = dS/S$, and $\alpha = dV/V$ as well as $\alpha = dn/n$. Substituting these variations into the fundamental relation for an open system, Equation (10.32), produces the relation

$$\mu = \frac{E - TS + PV}{n}.$$

(10.34)

In other words, the chemical potential of a system composed of a single substance is its Gibbs free energy, $G = E + TS - PV$, per unit mole. The same result can be derived from any other form of the fundamental constraint. For instance, if we impose the same variation on $dH = TdS + VdP + \mu dn$, including the constraint $dP = 0$ on the intensive variable P we again arrive at Equation (10.34).

The chemical potential of an ideal gas in terms of independent variables S, V, and n may be found from (10.34), the equations of state, and the function $E(S,V)$ taken from (9.13). The result is

$$\mu(S,V,n) = e_o \frac{\exp\left\{\dfrac{S - ns_o}{nc_V}\right\}}{\left(\dfrac{V}{nv_o}\right)^{R/c_V}}\left[1 - \frac{(S - nR)}{nc_V}\right],$$

(10.35)

where the dependence on n has been made explicit by defining the specific constants as $e_o = E_o/n$, $s_o = S_o/n$, $v_o = V_o/n$, and $c_V = C_V/n$.

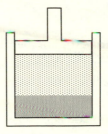

FIGURE 10.4 A cylinder and piston containing a system composed of two fluid phases.

Note that, as required, μ is an intensive variable. The chemical potential of other fluids and nonfluids may be found with a similar method.

We know that two systems in thermal equilibrium have the same temperature and that two fluid systems in mechanical equilibrium have the same pressure. These facts suggest that two open systems in equilibrium also have the same chemical potential. A general proof of this speculation requires developing general criteria for equilibrium, as we do in Chapter 12. Here I present a simpler, if more specialized, argument.

Consider two fluid phases, denoted 1 and 2, of one substance contained within the piston and cylinder diagramed in Figure 10.4. These phases could be a liquid in equilibrium with its vapor. Each phase can exchange mole number with the other. Therefore, their fluid variables are constrained by

$$dE_1 = T_1 dS_1 - P_1 dV_1 + \mu_1 dn_1 \tag{10.36}$$

and

$$dE_2 = T_2 dS_2 - P_2 dV_2 + \mu_2 dn_2. \tag{10.37}$$

Since the two phases are in thermal and mechanical equilibrium,

$$T_1 = T_2 = T \tag{10.38}$$

and

$$P_1 = P_2 = P. \tag{10.39}$$

Now, consider the properties of the larger system composed of both phases taken together. On the one hand, the composite system is a fluid described by the fluid variables E, T, S, P, and V. The composite system may exchange heat and work with its environment but not mole number. (Each phase exchanges mole number only with the other phase.) Therefore, its fundamental constraint is

$$dE = TdS - PdV. \tag{10.40}$$

On the other hand the system energy, apart from surface terms, which we assume to be negligible, is the sum of energies of the two phases, so that

$$E = E_1 + E_2. \tag{10.41}$$

In like manner the other extensive system variables are sums of their parts:

$$S = S_1 + S_2, \tag{10.42}$$

$$V = V_1 + V_2. \tag{10.43}$$

The sum of the phase mole numbers, however,

$$n = n_1 + n_2, \tag{10.44}$$

is a constant that enters into the description of the composite system not as a variable but only as a fixed parameter. For this reason, the moles gained by one phase are lost by the other—that is, $dn = 0$ or

$$dn_1 = -dn_2. \tag{10.45}$$

Adding the fundamental constraints of the two phases, (10.36) and (10.37), and using (10.38)–(10.39), (10.41)–(10.43), and (10.45) result in

$$dE = TdS - PdV + (\mu_1 - \mu_2)dn_1. \tag{10.46}$$

These two descriptions of the same composite system, (10.10) and (10.46), are consistent only when

$$\mu_1 = \mu_2. \tag{10.47}$$

Therefore, two phases are in equilibrium if and only if their temperature, pressure, and chemical potential are equal. (See Problems 10.8–10.10.)

10.6 Multivariate Systems

A fluid that exchanges mole number with its environment requires three independent variables for a complete description. So also does a gas composed of paramagnetic particles. The fundamental constraint of the former is $dE = TdS - PdV + \mu dn$, while that of the latter is $dE = TdS - PdV + (B_o/\mu_o)dB_m$. One can easily imagine more complex systems that require four independent variables—for example, a paramagnetic gas that exchanges mole number with its environment. In general, an independent variable is required for each way the system energy may be reversibly changed.

Equations of state, energy-characterizing functions, and cross-differentiations follow from a multivariable fundamental constraint in the same way these follow, in the two-variable case, from $dE = TdS - PdV$. As a relatively simple example we explore the consequences of the fundamental constraint,

$$dE = TdS - PdV + \mu dn, \tag{10.48}$$

of a fluid system that can exchange mole number with its environment. Its energy is a function, $E(S,V,n)$, of its three proper variables, S, V, and n. Its three equations of state take the form $T = (\partial E/\partial S)_{V,n}$, $-P = (\partial E/\partial V)_{S,n}$, and $\mu = (\partial E/\partial n)_{S,V}$. Following from these are three cross-differentiations

$$\left(\frac{\partial T}{\partial V} \right)_{S,n} = -\left(\frac{\partial P}{\partial S} \right)_{V,n}, \tag{10.49}$$

$$\left(\frac{\partial T}{\partial n}\right)_{S,V} = \left(\frac{\partial \mu}{\partial S}\right)_{n,V}, \tag{10.50}$$

and

$$-\left(\frac{\partial P}{\partial n}\right)_{S,V} = \left(\frac{\partial \mu}{\partial V}\right)_{S,n}. \tag{10.51}$$

The first, (10.49), reproduces a Maxwell relation already encountered [Eq. (8.7)], while (10.50) and (10.51) are unique to open fluid systems with variable mole number.

Legendre transformations produce old and new characterizing functions with the dimensions of energy E, $E + PV (\equiv H)$, $E - TS (\equiv A)$, $E - \mu n$, $E + PV - TS (\equiv G)$, $E + PV - \mu n$, $E - TS - \mu n$, and $E + PV - TS - \mu n$. Each of these has its own set of independent variables and each its own equations of state and cross-differentiations. Applying the reciprocity and reciprocal rules directly to (10.49)–(10.51) produces new cross-differentiations. (See Problems 10.11 and 10.12.)

Chapter 10 Problems

10.1 *Rubber Band Maxwell Relations.* Find the four Maxwell-type relations among first-order partial derivatives of rubber band variables T, S, F, and L implied by the fundamental constraint $dE = TdS + FdL$. Generate these with the reciprocity, reciprocal, and chain rules or from an appropriate transformation of the Maxwell relations for a fluid.

10.2 *Paramagnetic Material.* Show that the energy, E, of a paramagnetic material that obeys Curie's law is a function of temperature, T, alone. Work only from what we know about paramagnetic materials. (S)

10.3 *Magnetic Maxwell Relations.* Find the four Maxwell-type relations among first-order partial derivatives of the paramagnetic material variables T, S, B_o, and B_m implied by the fundamental constraint $dE = TdS + (B_o/\mu_o)dB_m$. Generate these with the reciprocity, reciprocal, and chain rules.

10.4 *Surface Work*. The surface tension of water, σ, is 0.073 N/m when $T = 298$ K. How much reversible work is required to spread a 3.0 mm-radius spherical drop of water into a circular sheet that is 50 cm in diameter? (Note: A sheet has two surfaces.)

10.5 *Constant Specific Heat of Surface*. Show that the characterizing energy $E(S,A) = cAT_c \exp(S/A)$ is consistent with the equations of state, $E = cAT$ and $\sigma = cT \ln(T_c/T)$, for a surface.

10.6 *Eötvös Equation*. The Eötvös equation of state for surface tension is $\sigma(T) = \sigma_o(1 - T/T_c)$, where $T \leq T_c$. Show that, given the Eötvös equation of state and the context of the theoretical picture given in the first part of Section 10.4,

    (a) the internal energy $E = \sigma_o A$, and

    (b) the entropy $S = \sigma_o A/T_c$. (S)

10.7 *Ferguson Equation*. The Ferguson equation of state for surface tension is $\sigma(T) = \sigma_o(1 - T/T_c)^n$, where $n \geq 1$ and is a generalization of the Eötvös equation of Problem 10.6. Assume the Ferguson equation of state and the theoretical picture given in the first part of Section 10.4. Find

    (a) the function $E(A,T)$, and

    (b) the function $S(A,T)$.

10.8 *Chemical Potential of an Ideal Gas, I*. Verify that Equation (10.35) follows directly from $\mu = (E - TS + PV)/n$ and what we know from Chapter 9 about the ideal gas.

10.9 *Chemical Potential of an Ideal Gas, II*. The chemical potential of an ideal gas, $\mu(S,V,n)$, may be derived directly from the partial derivative, $(\partial E/\partial n)_{S,V}$, of the energy of an ideal gas, $E(S,V,n)$, in which the system mole number enters as an explicit variable. To do so one must first rewrite $E(S,V)$ as found in Equation (9.13), that is,

$$\frac{E(S,V)}{E_o} = \frac{\exp\left\{\dfrac{S - S_o}{C_V}\right\}}{\left(\dfrac{V}{V_o}\right)^{nR/C_V}},$$

in terms of specific quantities $c_v = C_V/n$, $e_o = E_o/n$, $s_o = S_o/n$, and $v_o = V_o/n$ that are independent of the variables S, V, and n. Then show that integrating $\mu = (\partial E/\partial n)_{S,V}$ reproduces the result $\mu = G/n$ found in (10.35).

10.10 *Cavity Radiation.* Show that the chemical potential of cavity radiation is zero by showing that the Gibbs free energy of cavity radiation is zero. (That the Gibbs free energy of cavity radiation vanishes identically is a sign that the variables T and P of cavity radiation cannot both be independent variables.)

10.11 Start from the fundamental constraint $dE = TdS - PdV + \mu\, dn$ and the definitions of enthalpy, Helmholtz free energy, and Gibbs free energy and prove the following:

(a) $\mu = \left(\dfrac{\partial H}{\partial n}\right)_{S,P}.$

(b) $\mu = \left(\dfrac{\partial A}{\partial n}\right)_{T,V}.$

(c) $\mu = \left(\dfrac{\partial G}{\partial n}\right)_{T,P}.$

10.12 Prove the following relations among variables describing a single-phase, open fluid system.

(a) $\left(\dfrac{\partial T}{\partial n}\right)_{S,P} = \left(\dfrac{\partial \mu}{\partial S}\right)_{P,n}$ and $\left(\dfrac{\partial V}{\partial n}\right)_{S,P} = \left(\dfrac{\partial \mu}{\partial P}\right)_{S,n}.$ (S)

(b) $\left(\dfrac{\partial S}{\partial n}\right)_{T,V} = \left(\dfrac{\partial \mu}{\partial T}\right)_{V,n}$ and $\left(\dfrac{\partial P}{\partial n}\right)_{T,V} = \left(\dfrac{\partial \mu}{\partial V}\right)_{T,n}.$

(c) $\left(\dfrac{\partial S}{\partial n}\right)_{T,P} = \left(\dfrac{\partial \mu}{\partial T}\right)_{P,n}$ and $\left(\dfrac{\partial V}{\partial n}\right)_{T,P} = \left(\dfrac{\partial \mu}{\partial P}\right)_{T,n}.$

ELEVEN

Equilibrium and Stability

11.1 Mechanical and Thermal Systems

Recall that an equilibrium state is one described by a small set of thermodynamic variables that change only when the system's environment changes. Thus, two systems in continuous thermal contact are in mutual thermal equilibrium, and two systems in thermal equilibrium have, by definition, the same temperature. I have also referred to the concept of mechanical equilibrium, according to which two fluid systems in mutual mechanical equilibrium have the same pressure. This chapter gives these observations a new foundation.

When two systems completely isolated from the rest of the universe and initially isolated from one another are allowed to interact—for instance, through a heat-conducting wall, a movable partition, or a permeable membrane—they seek new equilibrium states. Not every equilibrium state is stable, however. The prediction of new stable equilibria is, according to H. B. Callen, the *fundamental problem of thermodynamics*.

Identifying the stable equilibria of a thermodynamic system is analogous to identifying the stable equilibria of a mechanical system. Figure 11.1 shows the potential energy function $U(x)$ of a mechanical system with one degree of freedom, here quantified with the variable x. Points where the first derivative of $U(x)$ vanishes, that is, where $dU/dx = 0$, are *equilibria*. These are further divided into *stable and unstable equilibria* according to the sign of the second derivative of the potential energy function d^2U/dx^2. Stable equilibria are those for which $dU/dx = 0$ and $d^2U/dx^2 > 0$; unstable equilibria are those for which $dU/dx = 0$ and $d^2U/dx^2 < 0$. A system displaced a very small distance from a point of stable equilibrium will oscillate around that point, while a system displaced a very small distance from an unstable equilibrium will be forced from its vicinity. Points B and D in Figure 11.1 are stable equilibria, while point C is an unstable equilibrium. Point A is not an equilibrium at all.

Thermodynamic systems differ from simple mechanical ones in having a very large number of internal degrees of freedom—on the order of several times the number of particles composing the system. Only when a thermodynamic system is in equilibrium are all these degrees of freedom effectively represented by a few thermodynamic variables. Thus, thermodynamic systems have many nonequilibrium states and can experience a variety of nonequilibrium processes.

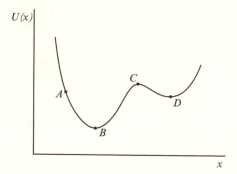

FIGURE 11.1   Potential energy function, $U(x)$, of a mechanical system with one degree of freedom. Points B, C, and D are equilibria; point A is not an equilibrium. Point B is most stable, C is unstable, and D is stable but less so than B.

Thermodynamic systems use nonequilibrium processes to seek out the most stable equilibrium state available consistent with the constraints imposed upon them. For this reason unstable equilibria are never maintained, and all states that are relatively but not absolutely stable have a limited lifetime. Of course, the latter so-called *metastable states* (analogous to point D in Fig. 11.1) may, if long-lived, be empirically indistinguishable from absolutely stable ones (analogous to point B).

Thermodynamic systems also differ from mechanical ones in the function minimized or maximized at stable equilibrium. The thermodynamic function of state maximized at stable equilibrium is, according to the entropy corollaries of Section 7.5, the entropy. From this idea we will develop intrinsic stability criteria for a fluid and a deeper understanding of why an initially single-phase fluid breaks into two phases.

11.2 Principle of Maximum Entropy

The criterion for identifying the stable equilibria of thermodynamic systems has already been encountered in Section 7.5. This criterion is the *principle of maximum entropy:*

> *A completely isolated system that may evolve only by diminishing its entropy is in stable equilibrium.*

Consider, for instance, the two systems illustrated in Figure 11.2. Rigid, adiabatic, and impermeable walls completely isolate the two

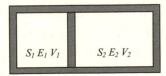

$S_1 E_1 V_1$ $\qquad$ $S_2 E_2 V_2$

FIGURE 11.2 Two systems completely isolated from each other and from the rest of the universe. When their mutual boundary is changed to one that conducts heat or is allowed to move, the two systems seek a new state of stable equilibrium.

systems from each other and from the rest of the universe. Their fundamental constraints are

$$dS_1 = \frac{1}{T_1} dE_1 + \frac{P_1}{T_1} dV_1 \tag{11.1}$$

and

$$dS_2 = \frac{1}{T_2} dE_2 + \frac{P_2}{T_2} dV_2. \tag{11.2}$$

As usual, $S_i = S_i(E_i, V_i)$, $1/T_i = (\partial S_i / \partial E_i)_{V_i}$, and $P_i/T_i = (\partial S_i / \partial V_i)_{E_i}$ where $i = 1, 2$. The entropy of the composite system is

$$S(E_1, V_1, E_2, V_2) = S_1(E_1, V_1) + (E_2, V_2). \tag{11.3}$$

When the two systems are completely isolated from one other, there is no particular relation between their temperatures, T_1 and T_2, and pressures, P_1 and P_2. If, however, we relax one of the constraints, the composite system will evolve to a new, stable equilibrium that maximizes its total entropy.

As an example, suppose we allow the two systems to interact thermally but not otherwise. In general, the systems will exchange heat irreversibly, and their energies, E_1 and E_2, will change subject to the constraints

$$E_1 + E_2 = E \text{ (a constant)} \tag{11.4}$$

and

$$V_1, V_2 = \text{constant.} \tag{11.5}$$

Constraint (11.4) allows us to eliminate the variable E_2 in favor of $E - E_1$. Then the entropy of the composite system reduces to a function of a single variable—that is, to $S(E_1) = S_1(E_1) + S_2(E - E_1)$, where we have suppressed dependence on the constants V_1 and V_2.

We are now in a position to apply the equilibrium ($dS/dE_1 = 0$) and stability ($d^2S/dE_1^2 < 0$) criteria associated with the principle of maximum entropy. In doing so, note that S_2 is a function of the variable E_1 only through the combination $E - E_1$. The chain rule produces

$$\frac{dS}{dE_1} = \frac{dS_1(E_1)}{dE_1} + \frac{dS_2(E - E_1)}{dE_1}$$

$$= \frac{dS_1(E_1)}{dE_1} + \frac{dS_2(E - E_1)}{d(E - E_1)} \cdot \frac{d(E - E_1)}{dE_1} \qquad (11.6)$$

$$= \frac{1}{T_1} - \frac{1}{T_2},$$

and so the equilibrium condition, $dS/dE_1 = 0$, reduces to

$$T_1 = T_2. \qquad (11.7)$$

The stability condition, $d^2S/dE_1^2 < 0$, follows from

$$\frac{d^2S}{dE_1^2} = \frac{d}{dE_1}\left(\frac{1}{T_1}\right) - \frac{d}{dE_1}\left(\frac{1}{T_2}\right), \qquad (11.8)$$

where $T_1 = T_1(E_1)$ and $T_2 = T_2(E_2) = T_2(E - E_1)$. The chain rule reduces (11.8) to

$$\frac{d^2S}{dE_1^2} = \left(\frac{-1}{T_1^2}\right)\frac{dT_1}{dE_1} + \left(\frac{1}{T_2^2}\right)\frac{dT_2}{dE_1}$$

$$= \frac{-1}{T_1^2 C_{V,1}} + \frac{-1}{T_2^2 C_{V,2}}, \qquad (11.9)$$

where $dE_i/dT_i = C_{V,i}$ because the V_i, for $i = 1, 2$, are constants. Evaluating d^2S/dE_1^2 at the equilibrium condition $T_1 = T_2$ further reduces (11.9) to

$$\frac{d^2S}{dE_1^2} = \left(\frac{-1}{T^2}\right)\left(\frac{1}{C_{V,1}} + \frac{1}{C_{V,2}}\right). \qquad (11.10)$$

Since temperatures are real numbers, $d^2S/dE_1^2 < 0$ is equivalent to

$$\left(\frac{1}{C_{V,1}} + \frac{1}{C_{V,2}}\right) > 0. \qquad (11.11)$$

Inequality (11.11) is not our final result, however. Since we have made no assumptions concerning the size or identity of systems 1 and 2 other than that they are both fluids in thermal equilibrium, condition (11.11) must hold generally. In particular, suppose system 2 is identical to system 1. Then $C_{V,1} = C_{V,2} [\equiv C_V]$, and (11.11) becomes

$$C_V > 0. \tag{11.12}$$

Inequality (11.12) expresses a condition of *intrinsic stability*. Any fluid in equilibrium must have a positive heat capacity at constant volume C_V.

Because the mathematical tools we have employed thus far depend only upon the local properties of the entropy function—that is, on its first and second derivatives at a point—they do not, in general, distinguish between relative and absolute stability. Yet, because there is only one point of stable equilibrium in this simple application, it must be absolutely stable. In similar ways, the context often allows us to distinguish between relative or metastable and absolutely stable equilibria.

11.3 Other Stability Criteria

All stability criteria originate from the differential Clausius inequality

$$dQ \le T_o dS, \tag{11.13}$$

where here, T_o is the temperature of the environment with which the system exchanges heat, dQ, and dS is the differential entropy change generated by this heat exchange. Only when the change associated with the variation is reversible is the equality sign realized and the temperature of the environment, T_o, equal to the temperature of the system, T. The Clausius inequality and all inequalities that follow from it constrain the allowed variations that result from equilibrium and nonequilibrium processes.

Imagine an indefinitely complex system that may contain different phases and kinds of particles or may be fragmented into different parts whose interactions are constrained in various ways. However complex, we allow this system to interact with its environment in only two ways: heat exchange, dQ, and work, PdV. Furthermore, we require its environment to be so large compared with the system as to be a reservoir with constant temperature T_o and pressure P_o. Since the following analysis is theoretical, we have no difficulty in imposing these and other constraints (for example, constant entropy) that would be difficult to realize in practice.

If the complex system is completely isolated from its environment, $dQ = 0$, $dV = 0$, and consequently $dE = 0$. Then the Clausius inequality (11.13) reduces to

$$(dS)_{E,V} \geq 0, \tag{11.14}$$

where the subscripts E and V denote the system state variables held constant during the variation dS. Inequality (11.14) allows only those differential variations that, leaving the system energy and volume unchanged, also leave unchanged or increase the system entropy. From here it is only a short step to the *principle of maximum entropy*:

> A completely isolated system that may evolve only by diminishing its entropy is in stable equilibrium.

We have already applied the principle of maximum entropy to a system composed of two homogeneous subsystems, but this derivation allows for much broader application.

By redefining the way this complex system may interact with its environment, we derive other stability criteria. Some of these will prove very convenient. We prepare for these derivations in the following way. The heat, dQ, absorbed by the system and work, $-P_o dV$, done on the system are related to each other by the first law, $dE = dQ - P_o dV$. This relation in the form $dQ = dE + P_o dV$ transforms the Clausius inequality (11.13) into

$$dE + P_o dV \leq T_o dS. \tag{11.15}$$

The principle of maximum entropy follows immediately from (11.15) but so also do other stability criteria.

Suppose we allow only those changes that preserve the system volume and entropy—that is, we adopt the constraints $dV = 0$ and $dS = 0$. Then (11.15) reduces to

$$(dE)_{V,S} \leq 0. \tag{11.16}$$

The constraints imposed (constant volume and entropy) prohibit variations that increase the energy. Thus, this variation establishes the following *principle of minimum energy*:

A system that, preserving its volume and entropy, can change only by increasing its energy is in stable equilibrium.

Next, suppose we allow only those changes that preserve the system entropy and maintain mechanical equilibrium with the reservoir so that $dS = 0$ and $dP = 0$. In this case (11.15) reduces to $[d(E + P_o V)]_{S,P} \leq 0$, which, given the condition of mechanical equilibrium $P = P_o$ and the definition of enthalpy $H = E + PV$, is equivalent to

$$(dH)_{S,P} \leq 0. \tag{11.17}$$

Since the constraints imposed prohibit those variations that increase the system enthalpy, we have the *principle of minimum enthalpy*:

A system that, preserving its entropy and maintaining constant pressure, can change only by increasing its enthalpy is in stable equilibrium.

Suppose we allow only those changes that preserve the system volume and maintain thermal equilibrium with the environment so that $dV = 0$ and $dT = 0$. Then (11.15) reduces to $[d(E - T_o S)]_{V,T} \leq 0$, which, given the equilibrium condition $T = T_o$ and the definition of Helmholtz free energy $A = E - TS$, is equivalent to

$$(dA)_{V,T} \leq 0. \tag{11.18}$$

These variations never increase the Helmholtz free energy and lead to the *principle of minimum Helmholtz free energy*:

> A system that, preserving its volume and maintaining constant temperature, can change only by increasing its Helmholtz free energy is in stable equilibrium.

Finally, allowing only those changes that maintain mechanical and thermal equilibrium with the reservoir, we have the constraints $dT = 0$ and $dP = 0$. In this case (11.15) reduces to $[d(E + P_o V - T_o S)]_{V,T}$ ≤ 0, which, given $T = T_o$, $P = P_o$, and the definition of Gibbs free energy $G = E + PV - TS$, is equivalent to

$$(dG)_{T,P} \leq 0. \tag{11.19}$$

These allowed variations never increase the Gibbs free energy and establish the *principle of minimum Gibbs free energy*:

> A system that, maintaining constant temperature and pressure, can change only by increasing its Gibbs free energy is in stable equilibrium.

Table 11.1 summarizes these allowed variations and the stability criteria that follow from them. Each derives from the Clausius inequality applied to a complex system that is interacting, or not interacting, with an environmental reservoir. Since a system's entropy is rarely controlled, the principles of minimum energy and enthalpy are less useful in practical applications than the others. The Gibbs free energy is especially important because the constraints, $dT = 0$

TABLE 11.1   Stability criteria for thermodynamic systems

Constraint	Constraint	Allowed variation	Equilibrium criterion	Stability criterion
$dE = 0$	$dV = 0$	$(dS)_{E,V} \geq 0$	$(dS)_{E,V} = 0$	S max
$dS = 0$	$dV = 0$	$(dE)_{S,V} \leq 0$	$(dE)_{S,V} = 0$	E min
$dS = 0$	$dP = 0$	$(dH)_{S,P} \leq 0$	$(dH)_{S,P} = 0$	H min
$dV = 0$	$dT = 0$	$(dA)_{V,T} \leq 0$	$(dA)_{V,T} = 0$	A min
$dT = 0$	$dP = 0$	$(dG)_{T,P} \leq 0$	$(dG)_{T,P} = 0$	G min

and $dP = 0$, that lead to its minimization are exactly those naturally imposed by maintaining thermal and mechanical equilibrium with the atmosphere.

11.4 Intrinsic Stability of a Fluid

The stability criteria described above and listed in Table 11.1 allow us to uncover the several conditions that make simple fluids intrinsically stable. We have already discovered, from the principle of maximum entropy, one such condition: a fluid in stable thermal equilibrium with itself must have $C_V > 0$. The other conditions are also very plausible, even expected. Still, we learn much by seeing how they emerge from the stability criteria.

We need only consider a composite system composed of two simple fluid subsystems, 1 and 2—just as considered in Section 11.2 and illustrated in Figure 11.2. However, we vary the constraints applied to the composite system as well as those applied by the boundary separating the two subsystems in order to bring into play different stability criteria.

Suppose, for instance, the composite system has a constant volume and constant entropy. Then, according to Table 11.1, its total energy

$$E(S_1,V_1,S_2,V_2) = E_1(S_1,V_1) + E_2(S_2,V_2), \tag{11.20}$$

subject to the two constraints

$$V_1 + V_2 = V \text{ (a constant)} \tag{11.21}$$

and

$$S_1 + S_2 = S \text{ (a constant)}, \tag{11.22}$$

is a minimum at stable equilibrium. The constraints (11.21) and (11.22) reduce the energy, E, of the composite system to a function of only two variables, S_1 and V_1, so that

$$E(S_1, V_1) = E_1(S_1, V_1) + E_2(S - S_1, V - V_1). \tag{11.23}$$

First, let's further restrict the individual subsystem entropies, S_1 and S_2, to be constant while allowing the subsystems to exchange volume by doing reversible work on one another. Then (11.23) becomes the single-variable function

$$E(V_1) = E_1(V_1) + E_2(V - V_1), \tag{11.24}$$

where we have suppressed dependence on the constant S_1. Recall that $P = -(\partial E/\partial V)_S$. Therefore, equilibrium, $dE/dV_1 = 0$, and stability, $d^2E/dV_1^2 > 0$, conditions reduce to

$$P_1 = P_2 \tag{11.25}$$

and

$$-\frac{dP_1}{dV_1} + \frac{dP_2}{dV_1} > 0, \tag{11.26}$$

respectively. Because $P_1 = P_1(V_1)$ and $P_2 = P_2(V - V_1)$ the chain rule reduces (11.26) to

$$-\frac{dP_1}{dV_1} - \frac{dP_2}{dV_2} > 0. \tag{11.27}$$

Since the subscripts "1" and "2" are arbitrarily assigned, and the stability condition (11.27) must obtain for subsystems of any size, and since the derivatives dP_1/dV_1 and dP_2/dV_2 assume constant values of S_1 and S_2, (11.27) is equivalent to

$$\left(\frac{\partial P}{\partial V}\right)_S < 0. \tag{11.28}$$

Thus, only fluids with positive adiabatic compressibility $\kappa_S [= -V^{-1}(\partial V/\partial P)_S]$ are intrinsically stable. Increasing the pressure on an adiabatically enclosed fluid must decrease its volume. If we begin again with Equations (11.20)–(11.22) and this time preserve the volume of each subsystem, allowing each to exchange heat reversibly with the other, and then follow the same pattern of calculation, we find the equilibrium, $T_1 = T_2$, and the intrinsic stability, $C_V > 0$, conditions already given by (11.7) and (11.12).

Applying, in the same way, the principle of minimum Helmholtz free energy to two subsystems with constant, identical temperatures and with constant total volume while allowing each to do reversible work on each other results in the intrinsic stability condition

$$\left(\frac{\partial P}{\partial V}\right)_T < 0. \tag{11.29}$$

And, similarly, the principle of minimum enthalpy leads to

$$C_P > 0. \tag{11.30}$$

These conditions of intrinsic fluid stability—$C_V > 0$, $C_P > 0$, $(\partial P/\partial V)_T < 0$, and $(\partial P/\partial V)_S < 0$—are somewhat redundant, for one can show that $(\partial P/\partial V)_S < 0$ and $C_P > 0$ if and only if $(\partial P/\partial V)_T < 0$ and $C_V > 0$.

The conditions of intrinsic stability are examples of an even more general principle called *Le Châtelier's principle*:

> *A system in stable equilibrium that experiences a change in its state variables initiates processes that tend to restore the system to equilibrium.*

For example, when the pressure exerted by the environment on a stable fluid increases by even a small amount, the fluid's volume must decrease and, since $(\partial P/\partial V)_T < 0$, cause the fluid pressure to increase. In this way the fluid is brought back into equilibrium with its environment. Any fluid not having this property would catastrophically explode or collapse. (See Problems 11.1–11.4.)

Chapter 11 Problems

11.1 *Another Stability Criterion.* C. J. Adkins lists a number of stability criteria, including one that constrains the system energy E and entropy

S to be constants. Identify the associated state variable and whether it is minimized or maximized.

11.2 *Room-Temperature Solid.* Show that a room-temperature elastic solid is stable by showing that its equations of state observe $(\partial P/\partial V)_T < 0$, $(\partial P/\partial V)_S < 0$, and $C_p > 0$. (Note: $C_V > 0$ is part of the model definition.)

11.3 *Cavity Radiation.* Show that the equations of state for cavity radiation are intrinsically stable by showing that $(\partial P/\partial V)_T < 0$, $(\partial P/\partial V)_S < 0$, $C_V > 0$, and $C_p > 0$.

11.4 *Equivalence.* Show that $(\partial P/\partial V)_S < 0$ and $C_p > 0$ follow from the conditions $(\partial P/\partial V)_T < 0$ and $C_V > 0$ for any fluid.

[Hint: Use Equation (8.47) relating C_p and C_V and a condition analogous to (8.44) that relates $(\partial P/\partial V)_T$ and $(\partial P/\partial V)_S$.]

TWELVE

Two-Phase Systems

....

12.1 Phase Diagrams

....

Water is the prime example of a substance that, within the range of commonly experienced pressures and temperatures, realizes all three phases: solid, liquid, and vapor. Each phase coexists with another in equilibrium at certain temperatures and pressures. Ice floating in water melts at $T = 0.0°C$ and $P = 1$ atm, and liquid water boils at $T = 100°C$ and $P = 1$ atm. All three phases of water coexist in equilibrium at its triple point ($T = 273.16$ K, $P = 6 \ 10^2$ Pa, $V/n = 18 \ 10^{-6}$ m³/mol). Diagrams illustrating these regions of thermodynamic variable space are called *phase diagrams*.

The phase diagram of water is, however, neither the easiest to measure nor the easiest to understand once measured. In fact, CO_2 and several other gases that can be liquefied at relatively comfortable temperatures yielded the first accurately mapped phase diagrams. These originated with the Irish chemist Thomas Andrews (1813–1885), who began a series of experiments in 1861 in which he isolated CO_2 gas in a glass tube fitted with a piston. By compressing, decompressing, heating, and cooling the CO_2 Andrews explored its phase diagram.

If one compresses CO_2 gas slowly enough to keep it in thermal equilibrium with its environment, its pressure at first increases with decreasing volume in rough agreement with Boyle's law, $P \propto 1/V$. Eventually, at several tens of atmospheres of pressure, the CO_2 gas (or vapor as it is called in this regime) begins to liquefy and collect in the bottom of the tube. Compression of the liquid-vapor mixture merely converts more vapor into liquid until complete liquefaction is achieved. At this point further compression of the liquid requires very high pressures. Several of Andrews' CO_2 isotherms are shown in Figure 12.1.

That the liquid condensed in the bottom of Andrews' tube rather than at its top is, of course, a consequence of gravity. Without gravity, or, equivalently, in free fall, the liquid would collect indifferently in several parts of the tube. Still, the volume of each phase would increase or decrease at the expense of the other during compression and expansion.

Several features of Andrews' diagram attract our attention. One is that the horizontal isotherm-isobars in the two-phase liquid-vapor region shrink with increasing temperature, to a point—the *critical point* on the 87.7°F isotherm. A *critical pressure* P_c (72.8 atm) and a *critical molar volume* V_c/n (94.2 10^{-6} m³/mol) as well a *critical temperature* T_c (87.7°F) identify the critical point. Each pure substance has its own liquid-vapor *phase transition* and unique critical point. Critical point data for several substances are shown in Table 12.1.

Liquid and vapor are said to *coexist* in the two-phase region, which is sometimes called the *vapor dome*. Along the vapor dome boundary, indicated with the dashed line in Andrews' phase diagram (Fig. 12.1), the liquid and vapor phases are each *saturated*. The saturated liquid and vapor curves and the critical isotherm neatly divide phase space into four parts:

1. Within the vapor dome, liquid and vapor phases coexist in equilibrium.

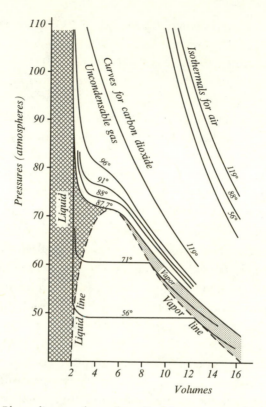

FIGURE 12.1 Phase diagram for carbon dioxide as measured by Thomas Andrews. Temperatures are in degrees Fahrenheit. Adapted from Mott-Smith, *The Concept of Heat,* p. 97.

2. Above its critical isotherm, no amount of pressure can liquefy a substance, which in this region is called a *gas.*

3. Below the critical isotherm and to the right of the saturated vapor curve, the substance is called a *vapor.*

4. Below the critical isotherm and to the left of the saturated liquid curve, the substance is liquid phase.

Each substance that boils at atmospheric pressure has a *normal boiling temperature,* defined as that temperature whose isotherm, within the vapor dome, falls on top of the $P = 1$ atm isobar. We

TABLE 12.1 Critical-point data for several substances

Substance	$\dfrac{P_c}{10^5\,Pa}$	$\dfrac{V_c}{n}\left(\dfrac{mol}{10^{-6}\,m^3}\right)$	$\dfrac{T_c}{K}$	$\dfrac{P_cV_c}{nRT_c}$
Ne	27	42	44	0.310
Ar	49	72	151	0.279
Kr	55	92	209	0.291
N_2	34	90	126	0.292
CO_2	74	94	304	0.275
H_2O	221	59	647	0.242
O_2	50	73	155	0.283
H_2	13	65	33	0.308

may experimentally control the boiling process as Andrews did by slowly and isothermally expanding a saturated liquid. During such expansion heat is drawn into the system and the system moves along an isotherm-isobar. In the kitchen we boil water in a less controlled fashion by heating the saturated liquid. Then the volume of the system expands rapidly. In principle, *boiling* takes place throughout the bulk of the whole liquid. *Evaporation*, on the other hand, occurs only on a liquid's surface.

Compressing a liquid drives it toward the liquid-solid phase transition. Within the two-phase region of this transition, liquid is converted into solid along an isotherm-isobar. This feature, illustrated in the generic phase diagram of Figure 12.2, recalls the liquid-vapor transition. But in many ways the two transitions are quite different. The solid-liquid transition has no associated critical temperature or critical point. And near room temperature, most solids expand only a few percent upon melting. In contrast, common liquids expand by a factor of a thousand or more upon vaporizing. Note that one can solidify but not liquefy a gas whose temperature remains above its critical temperature.

The right-hand boundary of the solid-liquid transition region meets the left-hand boundary of the liquid-vapor transition region at the *triple point*. Here, all three common phases of a substance coexist in equilibrium. Actually, the locus of points in *P-V-T* space at which vapor, liquid, and solid coexist forms a line, the *triple line*,

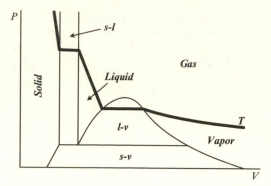

FIGURE 12.2 Isotherm (heavy line) in P-V space showing liquid-vapor and solid-liquid phase transitions. Solid-liquid (s-l), liquid-vapor (l-v), and solid-vapor (s-v) coexistence regions are indicated. Adapted from Walton, *Three Phases of Matter*, p. 12, Fig. 1.11c.

also shown in Figure 12.2. Below the triple line, a vapor and its solid coexist.

Phase diagrams may be represented with other pairs of coordinates or even as a surface extended in three-dimensional P-V-T space, as illustrated in Figure 12.3. James Clerk Maxwell (1831–1887) once created a plaster cast whose surface reproduced a phase diagram of water and shipped it across the Atlantic to the American thermodynamicist J. Willard Gibbs. Gibbs showed it to his Yale students but was quite reluctant to reveal the giver's identity—such was the fame of Maxwell and the modesty of Gibbs. Maxwell was fond of visualizations, but one need not go so far as to create a plaster cast. Merely acquiring the skill necessary to draw Figure 12.3 by hand would instill an intimate knowledge of the P-V-T surface. (See Problem 12.1.)

12.2 Van der Waals Equation of State

The structure of the vapor dome, with its parallel isotherms and isobars and critical point, cries out for a molecular-level explanation.

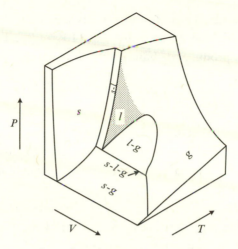

FIGURE 12.3 P-V-T surface of a substance that expands upon melting.

The Dutch physicist J. D. Van der Waals (1837–1923) provided such an explanation in 1879 and won the Nobel Prize for his work in 1910. Van der Waals reasoned that the assumptions that allow one to derive the ideal gas law, $PV = nRT$, from a molecular model fail for the liquid state. Chief among these failed assumptions are that molecules neither attract one another nor occupy space. Rather, Van der Waals knew that liquids cohere as if their molecules attract one another and strongly resist compression as if their molecules maintain a minimum spacing.

Thus, Van der Waals introduced a minimum volume, V_o, to the ideal gas law by subtracting V_o from V in $PV = nRT$. And he introduced molecular attraction by imagining that each molecule attracts all others, within a fixed range of influence, a fixed amount along a line connecting their centers. In this way, the net attractive force pulling any one molecule in a particular direction is proportional to the density of molecules, n/V, within its range of influence. Since the number of molecules that feel this pull is also proportional to n/V, the total attraction felt by the molecules on one side of a surface for molecules on the other side is proportional to the product of the net attraction exerted on each molecule and the density of molecules. Consequently, Van der Waals subtracted a term proportional to $(n/V)^2$ from the pressure, P, in $PV = nRT$. Together, these modifications produce the Van der Waals equation of state

$$P = \frac{nRT}{(V - nb)} - a\left(\frac{n}{V}\right)^2, \tag{12.1}$$

where the proportionality constant a and the minimum molar volume $b = V_o/n$ characterize a particular molecule. Figure 12.4 shows several Van der Waals isotherms that pass through and above the vapor dome.

We expect the energy equation of a Van der Waals fluid to be constrained but not completely determined by its equation of state (12.1). From the fundamental constraint $dE = TdS - PdV$ we have

$$\left(\frac{\partial E}{\partial V}\right)_T = T\left(\frac{\partial S}{\partial V}\right)_T - P, \tag{12.2}$$

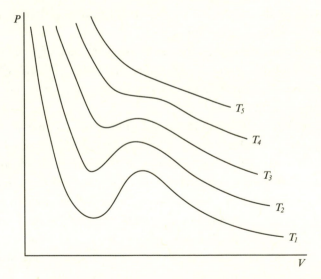

FIGURE 12.4 Isotherms of the Van der Waals equation of state. Adapted from Callen, *Thermodynamics*, p. 148, Fig. 9.1.

which, given the Maxwell relation $(\partial S/\partial V)_T = (\partial P/\partial T)_V$, is equivalent to

$$\left(\frac{\partial E}{\partial V}\right)_T = T\left(\frac{\partial P}{\partial T}\right)_V - P. \tag{12.3}$$

The right-hand side of Equation (12.3) can be evaluated on the basis of the Van der Waals equation of state (12.1) alone, and in this way, reduces (12.3) to the requirement

$$\left(\frac{\partial E}{\partial V}\right)_T = a\left(\frac{n}{V}\right)^2. \tag{12.4}$$

Integrating Equation (12.4) yields a form,

$$E = f(T) - a\frac{n^2}{V}, \tag{12.5}$$

consistent with a constant-volume heat capacity $C_V = df(T)/dT$. The Van der Waals equation of state and the laws of thermodynamics take us no further. If we assume a constant C_V, then (12.5) reduces to

$$E = C_V T - a\frac{n^2}{V}. \tag{12.6}$$

Note that equations (12.5) and (12.6) predict a finite temperature drop in a gas that iso-energetically increases its volume by expanding from one chamber into another without PdV work or heat exchange.

12.3 Two-Phase Transition

According to Figure 12.4 the Van der Waals isotherms have a positive slope within a certain region of the vapor dome. Yet we know that any fluid for which $(\partial P/\partial V)_T > 0$ is intrinsically unstable. The basic idea is, even apart from formal analysis, quite plausible: spatially separate parts of a constant-temperature fluid whose pressure increases as its volume increases (or whose pressure decreases as its volume decreases) would explode (or implode). Surrounding this region of intrinsic instability is a region of metastability. When a fluid is forced into either the unstable or the metastable region, it transitions to a nearby stable equilibrium state—an equilibrium state with lower Gibbs free energy and two distinct phases.

To illustrate this process let's map out how the pressure, $P(V,T)$, and Gibbs free energy, $G(P,T)$, evolve as a fluid moves along an isotherm from the liquid to the vapor side of its two-phase region— from A to B to C to D to E to F as shown in Figures 12.5a and 12.5b. Recall that $(\partial G/\partial P)_T = V > 0$. The topology of Figure 12.5b then follows inevitably from that of Figure 12.5a because the slope of the G versus P isotherm is positive and increases with increasing volume. As the fluid moves from A to B to C, the pressure and thus the Gibbs free energy decrease at a relatively slow rate. From point C to point D the slope $(\partial P/\partial V)_T$ changes sign so that $(\partial P/\partial V)_T > 0$, the system pressure increases with volume, and the Gibbs free energy increases at a rate proportional to its now larger volume. At point D the slope $(\partial P/\partial V)_T$ changes sign again so that $(\partial P/\partial V)_T < 0$, P resumes its decrease with increasing V, the volume V is yet larger, and the Gibbs free energy decreases more quickly. Note the point B-E in Figure 12.5b at which the Gibbs free energy curve necessarily intersects itself.

The dynamics of a two-phase system follows inevitably from the topology of Figures 12.5a and 12.5b. As the liquid is slowly expanded and its pressure decreased, its state evolves quasistatically along the isotherm from point A toward point B. Eventually the liquid crosses the intersection B-E and occupies a state just beyond B between B and C. Since there is a nearby state with less Gibbs free energy and the same temperature and pressure (found by following an isobar in Figure 12.5b down onto the curve FE), the principle of minimum Gibbs free energy ensures that this low Gibbs free energy state is sought out. In more physical terms, some liquid vaporizes and collects around bubbles that are usually present in the fluid while the remaining liquid remains at point B as saturated

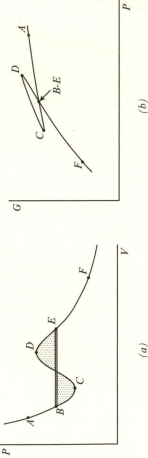

(a)

(b)

FIGURE 12.5 Isotherms of a two-phase system. (a) Isotherm in P-V space. States along path CD are intrinsically unstable, states along paths ED and CB are metastable, and states along paths AB and EF are stable. An isotherm-isobar (double line) connects points B and E. (b) Isotherm in G-P space. States along path CD are intrinsically unstable, states along paths BC and DE are metastable, and states along paths AB and EF are stable.

liquid. As the system is further expanded, more saturated liquid transforms into saturated vapor. Metastable states along BC and DE can be occupied only with great care—when an experimenter removes all bubbles and particles around which the fluid can vaporize and condense.

The points B and E, by construction, anchor the two ends of an isotherm-isobar, as shown in Figure 12.5a. The molar volume of the fluid at point B, $v_L \equiv V_L/n_L$, is that of the pure saturated liquid, and the molar volume at point E, $v_V \equiv V_V/n_V$, is that of the pure saturated vapor. In like manner the other extensive properties of the two coexisting phases are those whose molar values are associated with the phase points B and E. As the fluid vaporizes, a few moles of saturated liquid, dn_L, disappear from phase point B and, since the sum

$$n = n_L + n_V \tag{12.7}$$

is constant, reappear as moles of saturated vapor, $dn_V = -dn_L$, at phase point E. In this way, each of the two phases continues to occupy points B and E at either end of an isotherm-isobar on the boundary of the two-phase region. At the same time the volume of the two-phase system,

$$V = n_L v_L + n_V v_V, \tag{12.8}$$

evolves continuously from V_V to V_L. The two-phase system, as a whole, occupies all the intermediate points of the isotherm-isobar, while each phase, individually, occupies the ends of the isotherm-isobar.

The other extensive properties of a two-phase system evolve in similar fashion during condensation or vaporization. For example, the system energy

$$E = n_L e_L + n_V e_V, \tag{12.9}$$

entropy

$$S = n_L s_L + n_V s_V, \tag{12.10}$$

and heat capacity at constant volume

$$C_V = n_L c_V^L + n_V c_V^V \qquad (12.11)$$

are all continuous functions of n_L and n_V [$= n - n_L$]. In each case, lowercase letters, e_L, e_V, s_L, s_V, c_V^L, and c_V^V, denote molar quantities of the saturated fluid evaluated on the two-phase boundary (for example, $e_L \equiv E_L/n$, $e_V \equiv E_V/n$).

Figures 12.5a and 12.5b suggest a system of identifying phase transitions. As the fluid is compressed, the partial derivative $(\partial G/\partial P)_T$ changes discontinuously from V_V to $V_L < V_V$ at the transition point B-E. A phase transition accompanied by a discontinuity in the first-order partial derivatives of the Gibbs free energy, $(\partial G/\partial P)_T = V$ or $(\partial G/\partial T)_P = -S$, is a *first-order phase transition* according to a classification scheme invented by Paul Ehrenfest. Higher-order phase transitions—in which discontinuities first appear in higher-order partial derivatives of the Gibbs free energy [for example, in $(\partial^2 G/\partial P^2)_T$, $(\partial^2 G/\partial T^2)_P$, $(\partial^2 G/\partial P\partial T)$]—identify second- and higher-order phase transitions. (See Problems 12.2–12.4.)

12.4 Maxwell Construction

The saturated vapor and saturated liquid phase points lie on either end of an isotherm-isobar defined by a P-V-T equation of state. When this is a Van der Waals equation of state,

$$P = \frac{RT}{(V_L/n - b)} - a\left(\frac{n}{V_L}\right)^2 \qquad (12.12)$$

and

$$P = \frac{RT}{(V_V/n - b)} - a\left(\frac{n}{V_V}\right)^2. \qquad (12.13)$$

Together these two equations are not sufficient to determine the three functions $V_L(T)$, $V_V(T)$, and $P(T)$ that define the boundary of the two-phase region. However, we also know that saturated liquid and vapor phases on the same isotherm-isobar must have equal Gibbs free energies, $G(T,P)_L (\equiv E_L + PV_L - TS_L)$ and $G(T,P)_V (\equiv E_L + PV_L - TS_V)$. Thus, a third condition,

$$E_L + PV_L - TS_L = E_V + PV_V - TS_V, \tag{12.14}$$

obtains. The physics of this third condition (12.14) may also be expressed in terms of the variables P, V, and T, already appearing in the other two conditions. In this way we will, in principle, be able to solve the three equations—(12.12), (12.13), and the replacement of (12.14)—for the three unknowns functions, $V_L(T)$, $V_V(T)$, and $P(T)$, that define the boundary of the two-phase system.

In order to replace (12.14) with a more useful equivalent—known as the *Maxwell construction*—we prove that the work done by a two-phase system in expanding from 100% saturated liquid (phase point B in Fig. 12.5a) to 100% saturated vapor (phase point E) along the isotherm-isobar BE is the same as the work done by the system in expanding from B to E along the Van der Waals isotherm BCDE. That these two path integrals, $\int PdV$, are identical follows from the Helmholtz form of the fundamental constraint,

$$dA = -SdT - PdV, \tag{12.15}$$

and the requirement that $dT = 0$ along each path. Integrating (12.15) along the isotherm-isobar BE produces

$$A(T, V_V) - A(T, V_L) = - \int_{B \to E} PdV$$
$$= -P(T)(V_V - V_L). \tag{12.16}$$

On the other hand, integrating (12.15) along the Van der Waals isotherm from B to C to D to E produces

$$A(T, V_V) - A(T, V_L) = - \int_{V_L}^{V_V} P(V, T)\Big|_{T=const} dV, \tag{12.17}$$

where $P(V,T)$ is given by the Van der Waals P-V-T equation of state (12.1). Since the left-hand sides of equations (12.16) and (12.17) are identical, the condition

$$P(T)(V_V - V_L) = \int_{V_L}^{V_V} P(V,T)\Big|_{T=const} dV \qquad (12.18)$$

is graphically equivalent to choosing the isotherm-isobar in Figure 12.5a so that the two shaded areas are equal. Those states described by the equation of state but cut off from it by a properly chosen isotherm-isobar are, except with very special attention to suppressing instability, never realized.

For the Van der Waals equation of state, (12.1), condition (12.18) reduces to

$$P(T)(V_V - V_L) = \int_{V_L}^{V_V} \left[\frac{RT}{V/n - b)} - a\frac{n^2}{V^2} \right] dV. \qquad (12.19)$$

Although we can complete the integral in (12.19), the result and the other two conditions that define the vapor dome boundary, (12.12) and (12.13), must be solved numerically for the three functions $V_L(T)$, $V_V(T)$, and $P(T)$.

12.5 Clausius-Clapeyron Equation

The P-V-T equation of state of a fluid system with two independent state variables takes the form of an equation, $P = P(V,T)$, with two independent variables, here V and T, and a single dependent variable, here P. However, stable equilibria within a two-phase region are described by

$$P = P(T), \qquad (12.20)$$

where V assumes any value within a finite range defined by the boundary of the two-phase region. The function $P(T)$ also describes the projection of the two-phase region in *P-V-T* space, as shown in Figure 12.3, onto *P-T* space, as shown in Figure 12.6.

While the laws of thermodynamics alone cannot determine the function $P(T)$, Emile Clapeyron (1799–1864) discovered that the second law of thermodynamics does constrain the form taken by its derivative dP/dT, that is, constrains the slope of the curves in Figure 12.6. Clapeyron's analysis, published in 1832, was performed without benefit of the first law. Rudolph Clausius later built upon Clapeyron's result.

Recognizing that the Gibbs free energy, $G(T,P)$, is uniform along an isotherm-isobar within a two-phase region is the key to deriving the *Clausius-Clapeyron equation*. Consider the two differentially separated isotherm-isobars within the two-phase region of *P-V* space as shown in Figure 12.7. The Gibbs free energy of a two-phase system described with variables P, T, and volume V_a, here

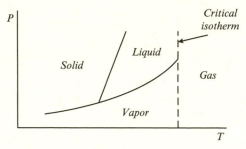

FIGURE 12.6 The two-phase regions of Fig. 12.3 projected onto curves in the *P-T* plane. The triple line (*s-l-g* in Fig. 12.3) projects onto a point.

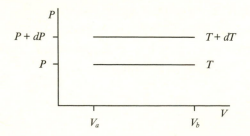

FIGURE 12.7 Differentially spaced isotherm-isobars in a two-phase region.

denoted $G(T,P)_a$, is identical in value to the Gibbs free energy with the same variables P, T, and a different volume V_b, that is, $G(T,P)_b$, since these points are on the same isotherm-isobar. Likewise, the Gibbs free energy with independent variables $P + dP$, $T + dT$, and volume V_a is identical in value to the Gibbs free energy with variables $P + dP$, $T + dT$, and volume V_b. Symbolically,

$$G(T,P)_a = G(T,P)_b \qquad (12.21)$$

and

$$G(T + dT, P + dP)_a = G(T + dT, P + dP)_b, \qquad (12.22)$$

where, of course, $G(T,P)_a \neq G(T + dT, P + dP)_a$ and $G(T,P)_b \neq G(T + dT, P + dP)_b$. Note, again, that the Gibbs free energy depends only on its independent variables, T and P, and not on the system volume, V. Rather, the volume V, according to the Gibbs form of the fundamental constraint $dG = -SdT + VdP$, specifies the derivative $(\partial G / \partial P)_T$.

Subtracting both sides of (12.21) from both sides of (12.22) produces

$$G(T + dT, P + dP)_a - G(T,P)_a = \qquad (12.23)$$
$$G(T + dT, P + dP)_b - G(T,P)_b,$$

which is equivalent to

$$dG(T,P)_a = dG(T,P)_b. \qquad (12.24)$$

Making use of the fundamental constraint $dG = -SdT + VdP$, Equation (12.24) becomes

$$-S_a dT + V_a dP = -S_b dT + V_b dP, \qquad (12.25)$$

that is,

$$\frac{dP}{dT} = \frac{(S_b - S_a)}{(V_b - V_a)}. \qquad (12.26)$$

The entropy difference $S_b - S_a$ is not directly measurable. However, $S_b - S_a$ is related to the latent heat of transition $Q_{a \to b}$ because

$$dQ_{rev} = TdS, \tag{12.27}$$

which can be integrated along an isotherm to yield

$$Q_{a \to b} = T(S_b - S_a). \tag{12.28}$$

This result reduces (12.26) to the Clausius-Clapeyron equation

$$\frac{dP}{dT} = \frac{Q_{a \to b}}{T(V_b - V_a)}, \tag{12.29}$$

where $Q_{a \to b}$ is the latent heat of transition of the process that takes the system from V_a to V_b along the two-phase isotherm-isobar.

The Clausius-Clapeyron equation determines the local slope of the P-T space phase-boundaries, as depicted in Figure 12.6, and explains several more or less commonly experienced phenomena. One is that the boiling temperature of water decreases with increased elevation and, consequently, decreased atmospheric pressure. For instance, at the boiling temperature of water, 100°C, and atmospheric pressure, $P = 1$ atm, the latent heat of vaporization of water is given by $Q_{L \to V} = 539$ cal/g, the volume of one gram of liquid water by $V_L = 1.04$ cm³/g, and the volume of one gram of water vapor by $V_V = 1.67 \ 10^3$ cm³/g so that $V_V - V_L \approx V_V$. These values imply

$$\frac{dP}{dT} = 3.62 \times 10^3 \cdot \frac{Pa}{°C}$$

$$= \frac{1}{28} \cdot \frac{atm}{°C}. \tag{12.30}$$

Thus, each 1/28 atm drop in pressure depresses the boiling point of water one degree Celsius. At the top of Mount Everest, where $P = 0.35$ atm, the boiling point of water is reduced by 0.65×28°C, or about 18°C—that is, to 82°C, provided that the boiling point depression rate remains constant over this interval.

Except near the critical point, vaporization increases the volume of a substance by a large factor. And vapor states at least roughly approximate an ideal gas. These approximations, $V_V - V_L \approx V_V$ and $V_V \approx nRT/P$, transform the Clausius-Clapeyron equation (12.29) into a useful rule:

$$\frac{dP}{dT} \approx \frac{Q_{L \to V}}{TV_V}$$

$$\approx \frac{(Q_{L \to V}/n)P}{RT^2}.$$

(12.31)

The freezing-point depression of water is a less directly experienced but also interesting phenomenon. At $T = 0$°C water contracts upon melting, from $V_S = 1.09$ cm^3/g to $V_L = 1.00$ cm^3/g so that $V_L - V_S = -0.09$ cm^3/g. Furthermore, the latent heat of fusion of water $Q_{S \to L} = 80$ cal/g. These values imply

$$\frac{dP}{dT} = -130 \cdot \frac{atm}{°C}.$$

(12.32)

Therefore, an increase of 130 atm depresses the freezing point—that is, the melting point—of water by one degree Celsius.

It has been claimed (and disputed) that the tiny edges of ice skate blades, as they support a human body, produce pressure sufficient to melt the ice below and allow the skater to glide on a layer of water. An even more dramatic demonstration of the freezing-point depression of water occurs when two weights are hung on the ends of a wire that passes over a block of ice. The force exerted by the wire lowers the melting point of the ice beneath the wire and, if the weights are sufficiently heavy, causes it to melt. The wire pushes through the water which then returns to atmospheric pressure and refreezes. In this way the wire gradually passes through an otherwise intact block of ice. (See Problems 12.5–12.7.)

12.6 Critical Point

The highest-temperature isotherm that passes through the two-phase region does so at the critical point where it *inflects*, that is, simultaneously satisfies,

$$\left(\frac{\partial P}{\partial V}\right)_T = 0 \tag{12.33}$$

and

$$\left(\frac{\partial^2 P}{\partial V^2}\right)_T = 0. \tag{12.34}$$

The Van der Waals equation of state (12.1) and the conditions (12.33), and (12.34) produce the critical point coordinates,

$$P_c = \frac{a}{27b^2}, \tag{12.35}$$

$$\frac{V_c}{n} = 3b, \tag{12.36}$$

and

$$T_c = \frac{8a}{27Rb}, \tag{12.37}$$

in terms of the parameters, a and b, that characterize the Van der Waals fluid. These, in turn, imply a Van der Waals *critical compression factor*

$$\frac{P_c V_c}{nRT_c} = \frac{3}{8} = 0.375 \tag{12.38}$$

that is about 30% higher than those found in Table 12.1. A difference of this size between Van der Waals theory and actual experimental outcomes is typical. (See Problems 12.8–12.11.)

Chapter 12 Problems

....

12.1 *P-V Phase Diagram.* Draw the *P-V* phase diagram for a substance that shows all three common phases. Label the solid, liquid, and gas phases; the various coexistence regions; the critical point; the triple point; and the triple line. Distinguish between gas and vapor.

12.2 *Lever Rule.* The lever rule relates the fraction in each phase of a two-phase system to its volume, V, relative to parameters at each end of the isotherm-isobar. The relation is that of a lever with "masses" n_L and n_V balanced at distances $(V - V_L)$ and $(V_V - V)$, respectively, from a pivot at V. Thus,

$$n_L (V - V_L) = n_V (V_V - V).$$

Derive the lever rule from the idea, encapsulated in Equations (12.7) and (12.8), that each phase coexisting in a two-phase system is the phase of a purely saturated fluid.

12.3 *Two-Phase Region.* One hundred grams of H_2O are at atmospheric pressure and 100°C. If 5 g are in vapor phase and 95 g in liquid phase, what is the total volume of the system? Use the specific volumes found near the end of Section 12.5.

12.4 *Heat Capacity.* Show that during a first-order phase transition in which $(\partial G/\partial T)_P$ changes discontinuously, the specific heat at constant volume $C_V \to \infty$.

12.5 *Boiling Point Elevation.* A particular liquid boils at 130°C at a pressure of 1.0 atm and has a heat of vaporization of 1.44×10^3 cal/mol. At what temperature will this liquid boil if its pressure is raised to 1.05 atm? Assume that dP/dT remains constant over this interval.

12.6 *Clausius-Clapeyron Rederived.* Derive the Clausius-Clapeyron equation (12.26) by applying the reciprocity theorem in the two-phase region. Follow these steps: (a) reciprocity on variables P, V, and T; (b) a Maxwell relation that shifts from variables P, V, and T to variables S, V, and T; (c) the chain rule; and finally, (d) the evaluation of partial derivatives in the two-phase region.

12.7 *Saturated Vapor Model.* Assume that the volume of a saturated liquid is ignorably small compared with the volume of the saturated vapor at the same temperature and that the ideal gas equation of state, $PV = nRT$, describes the saturated vapor.

(a) Given these assumptions derive an expression for $P(T)$ within a region of the vapor dome for which the heat of transition $Q_{l \to v}$ is a constant independent of temperature. Use the initial condition $P = P_o$ when $T = T_o$ to evaluate the integration constant.
[Hint: Use Equation 12.31.] (S)

(b) Use this relation and the data presented near the end of Section 12.5 to reestimate the number of degrees the boiling point of water is depressed at the top of Mount Everest.

12.8 *Van der Waals Critical Point.* Derive the Van der Waals critical point expressions $P_c = a/27b^2$, $V_c = 3bn$, and $T_c = 8a/27Rb$ from the conditions for inflection of the Van der Waals isotherm.

12.9 *Dieterici Equation of State.* The Dieterici equation of state,

$$P = \frac{RT}{\left(\dfrac{V}{n} - b'\right)} \exp\left\{\frac{-a'n}{RTV}\right\},$$

also models the liquid-vapor two-phase region. Find the critical point data, P_c, V_c, T_c, and the critical compression factor, $P_c V_c / nRT_c$, it implies. Is this $P_c V_c / nRT_c$ closer to those collected in Table 12.1 than that predicted by the Van der Waals equation of state?

12.10 *Mie-Grüneisen Equation of State.* The Mie-Grüneisen P-V-T equation of state,

$$P = \frac{n_m RT}{V} + \frac{3B_0}{(n - m)}\left[\left(\frac{V_0}{V}\right)^{(n+3)/3} - \left(\frac{V_0}{V}\right)^{(m+3)/3}\right],$$

where n_m is the number of moles, models the transition between a vapor and a generic condensed phase. With four characterizing parameters (n, m,

V_o, and B_o with $n > m$) it can fit data more closely than either the Van der Waals or the Dieterici P-V-T equation of state, since these have only two characterizing parameters. Find expressions for the critical point data P_c, V_c, T_c, and for the critical compression factor $P_c V_c / nRT_c$ in terms of the model's characterizing parameters.

12.11 *Cross-Differentiation.* Suppose we postulate the existence of equations of state having the forms $P = (nRT/V) + f(V)$ and $E = \int C_V(T)dT + g(V)$. Show that cross-differentiation arising from the entropy form of the fundamental constraint $dS = (1/T)dE + (P/T)dV$ requires that $f(V) = -dg(V)/dV$.

The Third Law

13.1 The Principle of Thomsen and Berthelot

The third law of thermodynamics was developed over a number of years, beginning in 1906 with the contributions of the German physical chemist Walther Nernst (1864–1941). Nernst proposed two different versions of what he called a new heat theorem. Max Planck (1858–1947) proposed a third, stronger version of the third law. While closely related, these three versions—here referred to as *entropy change*, *unattainability*, and *absolute entropy*—have distinct contents. Each is an attempt to bring the low-temperature behavior of physical systems under one principle.

The third law of thermodynamics is sometimes called the *Nernst postulate* after its originator. The designation *third law*, although very common, is in some ways pretentious. The third law is neither as foundational nor as consequential as the other laws of thermodynamics. Yet it cannot be derived from the other laws, and it provides a useful constraint on low-temperature experiments and theories.

Nernst's initial motivation was to explain the *principle of Thomsen and Berthelot*—actually more a rule of thumb than a principle. According to Julius Thomsen (1826–1909) and Marcellin Berthelot (1827–1907), chemical reactants tend to realize the most exothermic reaction possible, that is, the reaction that produces the most heat. Recall from Equation (8.18) that the heat absorbed, Q, by a constant-pressure system is the change in the system enthalpy, ΔH. Thus, according to the principle of Thomsen and Berthelot, chemical systems realize the state with the smallest possible enthalpy. This principle seemed a plausible explanation of many phenomena even if it was known, in the late nineteenth century, to have a number of exceptions.

But Nernst knew that the appropriate stability criterion for chemical systems maintained at constant temperature and atmospheric pressure is minimum Gibbs free energy, not minimum enthalpy. Gibbs free energy, $G[\equiv E + PV - TS]$, is related to enthalpy, $H[\equiv E + PV]$, through their definitions by

$$H = G + TS. \tag{13.1}$$

Thus, changes in the enthalpy, ΔH, and Gibbs free energy, ΔG, in a constant-temperature process are related by

$$\Delta H = \Delta G + T\Delta S. \tag{13.2}$$

That, according to (13.2), $\Delta H = \Delta G$ in the limit $T \to 0$ alone does not explain the relative usefulness of the principle of Thomsen and Berthelot. But if ΔS were to remain small as $T \to 0$, the principle of Thomsen and Berthelot might hold over a considerable range of temperatures.

13.2 Entropy Change

Nernst adopted the limit $\Delta S \to 0$ as $T \to 0$ as the entropy change version of the third law:

The entropy change ΔS in any reversible isothermal process approaches zero as the temperature approaches zero.

Entropy change implies that certain thermodynamic coefficients vanish as $T \to 0$. Consider, for instance, a fluid described in terms of the independent variables T and V, as is appropriate when adopting the Helmholtz form of the fundamental constraint, $dA = -S dT - P dV$. Then the equations of state assume the form $S = S(V,T)$ and $P = P(V,T)$. Because the partial derivative $(\partial S/\partial V)_T$ is defined by the limit

$$\left(\frac{\partial S}{\partial V}\right)_T = \lim_{\Delta V \to 0} \left[\frac{S(V + \Delta V, T) - S(V,T)}{\Delta V}\right]_{T=cons} \tag{13.3}$$

and, according to entropy change $\Delta S = S(V + \Delta V, T) - S(V,T) \to 0$ as $T \to 0$, then

$$\left(\frac{\partial S}{\partial V}\right)_T \to 0 \text{ as } T \to 0 \tag{13.4}$$

as long as these two limits, $\Delta V \to 0$ and $T \to 0$, may be taken in either order. For the same reason

$$\left(\frac{\partial S}{\partial P}\right)_T \to 0 \text{ as } T \to 0. \tag{13.5}$$

These limits, in turn, imply, through Maxwell relations (8.25) and (8.26), that

$$\left(\frac{\partial V}{\partial T}\right)_P \to 0 \text{ as } T \to 0 \tag{13.6}$$

and

$$\left(\frac{\partial P}{\partial T}\right)_V \to 0 \text{ as } T \to 0. \tag{13.7}$$

Thus, according to the entropy change version of the third law, the isobaric expansivity $\alpha_P = V^{-1}(\partial V/\partial T)_P$ and pressure coefficient $\alpha_V = P^{-1}(\partial P/\partial T)_V$ both vanish in the limit $T \to 0$. Furthermore, according to (8.46), $C_P - C_V = T(\partial V/\partial T)_P(\partial P/\partial T)_V$. Therefore, entropy change also requires that

$$C_P - C_V \to 0 \text{ as } T \to 0. \tag{13.8}$$

That these predictions have been frequently confirmed supports the entropy change version of the third law.

Further deductions from entropy change require adopting the *finite entropy hypothesis*—that the entropy $S(V,T)$ remains finite as V is held constant and $T \to 0$. This hypothesis is natural to and explicit in the absolute entropy version of the third law that we consider later. Here we adopt the finite entropy hypothesis as a separate assumption. According to entropy change,

$$\Delta S = S(V_2,T) - S(V_1,T) \to 0 \text{ as } T \to 0 \tag{13.9}$$

for arbitrary V_2 and V_1. Thus, we may replace V_2 in (13.9) with V_3 or any other value of V and, using the finite entropy hypothesis, conclude that

$$S(V_1,0) = S(V_2,0) = S(V_3,0) \ldots, \tag{13.10}$$

where the notation $S(V,0)$ denotes the limit of $S(V,T)$ as $T \to 0$. This means that all curves $S = S(V,T)$ for constant V must approach a common value of entropy $S_o \equiv S(V,0)$ as $T \to 0$ for arbitrary V, that is, each system approaches a unique entropy as $T \to 0$. Additionally, all isochors must have positive slope $(\partial S/\partial T)_V > 0$ when $T > 0$, since $C_V = T(\partial S/\partial T)_V$ and $C_V > 0$ is required of any $T > 0$ system in stable equilibrium. These properties are illustrated in Figure 13.1.

13.3 Unattainability

The unattainability version of the third law follows from and is logically equivalent to the entropy change version of the third law. According to unattainability, the equivalent of which was first proposed by Nernst in 1912,

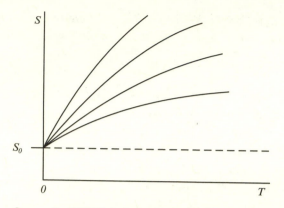

FIGURE 13.1 Curves $S = S(V,T)$ for constant values of V. According to the entropy change version of the third law and the finite entropy hypothesis, these isochors must all approach the same entropy, S_o, as $T \to 0$. Stability requires that $(\partial S/\partial T)_V > 0$ when $T > 0$.

> *No reversible adiabatic process starting at a non-zero tempera-*
> *ture can take a system to zero temperature.*

We cannot, for instance, cool a gas to zero temperature by reversibly and adiabatically (that is, isentropically) expanding the gas. Unattainability also applies to the technique of adiabatic demagnetization. We cannot cool a magnetic material to zero temperature by reversibly and adiabatically turning down the applied magnetic field.

In order to prove the equivalence of entropy change and unattainability we need to prove two propositions: (1) that entropy change implies unattainability and (2) that unattainability implies entropy change. Equivalently, and more conveniently, we prove the contrapositive of each of these propositions. In these proofs we take for granted the finite entropy hypothesis that $S(V,T)$ remains finite as $T \to 0$ and the stability requirement that $(\partial S/\partial T)_V > 0$ whenever $T > 0$.

The contrapositive of proposition 1 is that a denial of unattainability leads to a denial of entropy change. To deny unattainability

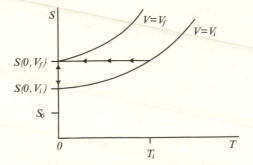

FIGURE 13.2 Two impossible isochors, $V = V_i$ and $V = V_f$, that define an isentrope intersecting the $T = 0$ axis and a finite isotherm along the $T = 0$ axis. These are used in proving the logical equivalence of the entropy change and unattainability versions of the third law.

means that an isentropic process exists that intersects the $T = 0$ axis. Such a process would take the system along an isentrope from an initial state with $V = V_i$ and $T = T_i$ to a final state with $V = V_f$ and $T = 0$, as illustrated in Figure 13.2. Isochors that pass through the endpoints of this isentrope are also shown. These isochors must intersect the $T = 0$ axis at different values of entropy, $S(V_i, 0)$ and $S(V_f, 0) \neq S(V_i, 0)$, as shown. Consequently, $S(V_f, T) - S(V_i, T)$ does not vanish in the limit $T \to 0$ as required by entropy change. Therefore, a denial of unattainability leads to a denial of entropy change.

The contrapositive of proposition 2 is that a denial of entropy change leads to a denial of unattainability. If we deny entropy change, an isothermal process $V_i \to V_f$ exists for which the entropy change $S(V_f, T) - S(V_i, T)$ does not approach zero as $T \to 0$. This supposed nonvanishing, zero-temperature, isothermal process is also illustrated in Figure 13.2. Consequently, isochors, $V = V_i$ and $V = V_f$, emerge from each end of this isotherm and further define an isentrope that intersects the $T = 0$ axis at the point $S = S(V_f, 0)$. In this way unattainability is denied. Therefore, denial of entropy change leads to a denial of unattainability, and the logical equivalence of these two versions of the third law has been demonstrated.

The unattainability version of the third law is often loosely characterized as declaring impossible all processes that attain absolute zero temperature. Certainly, unattainability prohibits any process from reaching the point $T = 0$ and $S = S_0$ in a finite number of isotherms and isentropes. Such process would have to include at least one isentrope that intersects the $T = 0$ axis, and this is explicitly forbidden by the unattainability version of the third law. Thus, any process that begins with finite-sized isotherms and isentropes and attempts to angle its way toward the points $T = 0$ and $S(V_f, T)$ $- S(V_i, T)$ without intercepting either the $T = 0$ or the $S = S_0$ axes, as illustrated in Figure 13.3b, must eventually fail or become a smooth path. However, an equilibrium process that follows a smooth path in S-T space, as illustrated in Figure 13.3a, is not forbidden by unattainability, since this path is effectively composed of an infinite number of isotherms and isentropes. Neither does unattainability forbid a non-equilibrium process that somehow takes a fluid system from an arbitrary state to the $T = 0$ and $S = S_o$ state. However, it is difficult to imagine how such a smooth-path or a non-equilibrium process might, in practice, be realized.

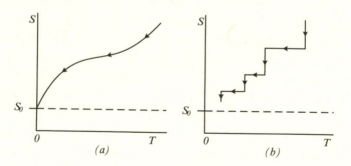

FIGURE 13.3 (a) A supposed smooth-curve reversible process that reaches the $T = 0$, $S = S_o$ state. (b) A process composed of isotherms and isentropes that attempts to reach the $T = 0$, $S = S_o$ state without first intersecting either the $T = 0$ or the $S = S_o$ axes.

The finite entropy hypothesis plus the entropy change (or unattainability) version of the third law together constitute the absolute entropy version of the third law. This version of the third law originates with Max Planck, whose own description is worth quoting: "The gist of the theorem is contained in the statement that, as the temperature diminishes indefinitely, the entropy of a chemically homogeneous body of finite density approaches indefinitely near to a definite value, which is independent of the pressure, the state of aggregation and the special chemical modification."

Note the two parts of Planck's statement: (1) that the entropy of a system approaches a definite (that is, finite) value as $T \to 0$ and (2) that this limiting value is independent of other thermodynamic variables, including chemical modifications to the system. Part 2 is equivalent to Nernst's entropy change version, while part 1 is the finite entropy hypothesis. Thus, absolute entropy is a stronger statement than either entropy change or unattainability per se; the former implies the latter, but the latter do not imply the former.

One consequence of the restriction to finite limiting entropies is that the heat capacity

$$C_V \to 0 \text{ as } T \to 0 \tag{13.11}$$

and, in view of (13.8),

$$C_p \to 0 \text{ as } T \to 0. \tag{13.12}$$

To demonstrate (13.11), integrate the relation $C_V = T(\partial S/\partial T)_V$ and so produce

$$S(V,T) - S(V,0) = \int_0^T \frac{C_V(V,T')}{T'} \, dT'. \tag{13.13}$$

Given that $S(V,0)$ and $S(V,T)$ are finite, the integral on the right-hand side of (13.13) must be finite, its integrand must be integrable, and, consequently, $C_V \to 0$ as $T \to 0$. Calorimetric measurements have confirmed the zero-temperature limits, (13.11) and (13.12), of the heat capacities C_V and C_P.

Evidence that the entropy of a system approaches the same finite value as $T \rightarrow 0$ independently of its volume, pressure, or arrangement of atoms into molecules requires comparing the entropies of a system in different states as each approaches zero thermodynamic temperature. Since only the entropy difference between two states connected with a reversible process can be measured, such comparisons require the reversible transformation of one of two initially identical systems into another state and the cooling of each system to very low temperatures. The transformation could be the chemical decomposition of one substance into component substances, a structural (that is, allotropic) transition from one form of solid into another, or a transition from liquid to glass or glass to liquid. In any case, the entropy change attending the transition of one system into another state and the subsequent cooling is compared with the entropy change attending the cooling of the originally identical system. The absolute entropy and entropy change forms of the third law require that these entropy changes be identical, and careful measurements show that they are.

Interestingly, the absolute entropy version of the third law makes no claim about the relative value of the limiting entropy of systems that cannot be reversibly transformed into one another. In fact, the entropy of systems that cannot be reversibly transformed into one another cannot be compared with one another. Such thermodynamically disconnected systems include those composed of nuclei with different neutron-to-proton ratios. However, this inability to compare gives us the freedom to adopt as a convention the same (or even differing) limiting entropies for thermodynamically disconnected systems. Those who construct tables of absolute entropy for a variety of elements take advantage of this freedom, although they do not often mention that the relative size of entropies belonging to systems that cannot be reversibly transformed into each other carries no information.

Occasionally one finds formulations of the third law that miss this, possibly subtle, point. For instance, the statement that "the entropy of all perfect crystals is the same at thermodynamic zero" mixes physical law and allowed convention. A more correct version of the third law using the same language would be "the entropy of all perfect crystals may be taken to be the same at thermodynamic zero."

Chapter 13 Problem

13.1 *Low-Temperature Limit of Cavity Radiation.* Many of the equations of state discussed in this volume incorporate a constant heat capacity C_V appropriate only for high-temperature systems. However, the equations of state for cavity radiation, in Section 9.3, should obtain for all temperatures, including those down to the $T \to 0$ limit. Show that cavity radiation observes the following limits.

(a) $\left(\dfrac{\partial S}{\partial V} \right)_T \to 0$ as $T \to 0$, (S)

(b) $\left(\dfrac{\partial P}{\partial T} \right)_V \to 0$ as $T \to 0$,

(c) $C_V \to 0$ as $T \to 0$.

Appendix A

....

Physical Constants and Standard Definitions

Physical Constants in SI Units

Name	Symbol and Value
Name	*Symbol and Value*
Gas constant	$R = 8.31$ J/(K mole)
Stefan-Boltzmann constant	$\sigma_B = 5.67 \times 10^{-8}$ W/(m^2 K^4)
Magnitude of the electric charge	$e = 160 \times 10^{-19}$ C
Acceleration of gravity	$g = 9.8$ m/s^2
Speed of light in vacuum	$c = 3.00 \times 10^8$ m/s
Radiation constant	$a = 4\sigma_B/c = 7.56 \times 10^{-16}$ kg/(s^2 K^{4}m)
Permeability of free space	$\mu_o = 4\pi \times 10^{-7}$ H/m
Mechanical equivalent of heat	$J = 4.186$ J/cal

Standard Definitions

Name	Symbol and Definition
Name	*Symbol and Definition*
Atmospheric pressure	1 atm $= 1.01 \times 10^5$ Pa
	$= 760$ Torr
	$= 760$ mm Hg
	$= 14.7$ lbs/in^2
Bar	1 bar $= 10^5$ Pa
Standard temperature	273 K
Standard (or normal) pressure	1 atm
Volume of ideal gas at standard temperature and pressure	22.4 m^3/mol
Nutritional calorie	1 Cal $= 10^3$ cal
Liter	1 L $= 10^3$ cm$^3 = 10^{-3}$ m^3

Appendix B

··

Catalog of 21 Simple Cycles

One way to characterize the logical content of a statement is to identify all of its consequences. But the first and second laws of thermodynamics are too fundamental for this kind of characterization; they have far too many consequences. Yet, we can do the next best thing: identify all the consequences of the laws for a certain small but important class of cycles—those cyclic processes that exchange heat with two or fewer heat reservoirs and that may or may not produce or consume work. There are 21 distinct simple cyclic processes belonging to this class. Indefinitely complex cycles can be built up out of these simple ones. Each of the 21 simple cycles is diagrammed in Figure B.1. Note that when heat is exchanged with only one reservoir the distinction between hotter and colder reservoirs collapses.

Three of the 21 catalog cycles, identical to those in Figure 5.4, are explicitly allowed. These are already marked, in Figure B.1, with the symbol ✓. The first law and each version of the second law prohibit certain other simple processes. Our task is to find the other cycles necessarily forbidden by the first law or a particular version of the second law and mark them with the symbol ⊘. We assume that all those not forbidden are allowed. In this way each thermodynamic law will generate its own pattern of ✓ and ⊘ symbols on the catalog—a kind of logical fingerprint. Those versions of the second law with identical fingerprints are logically equivalent.

By "necessarily violate" the first law I mean cycles that can be proven equivalent to one of the one-arrow cycles, diagrammed in Figure 4.3. These directly violate the first law. By "necessarily violate" a version of the second law I mean cycles that can be proven equivalent to one of the cycles, diagrammed in Figures 5.2 and 5.3,

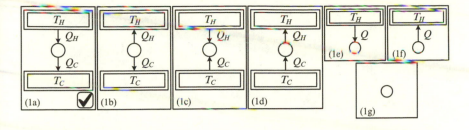

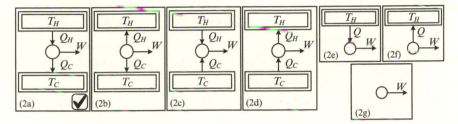

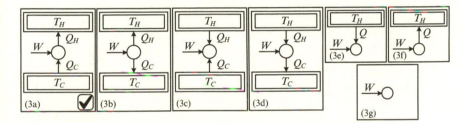

FIGURE B.1 The catalog of 21 simple cycles that exchange heat with two or fewer heat reservoirs and that may or may not produce or consume work. Each one of the explicitly allowed cycles, also displayed in Fig. 5.4, is here marked with the symbol ✓.

that directly violates a version of the second law. Catalog cycles that necessarily violate the first law of thermodynamics are fairly obvious: (1b), (1c), (1e), (1f), (2b), (2f), (2g), (3c), (3e), and (3g). These are marked ⊘ in Figure B.2. Others, such as (1d), do not necessarily violate the first law because their violation or nonviolation is contingent upon the values of Q_H, Q_C, and W.

When inspecting the consequences of a particular version of the second law, say, Thomson's second law, one can immediately mark

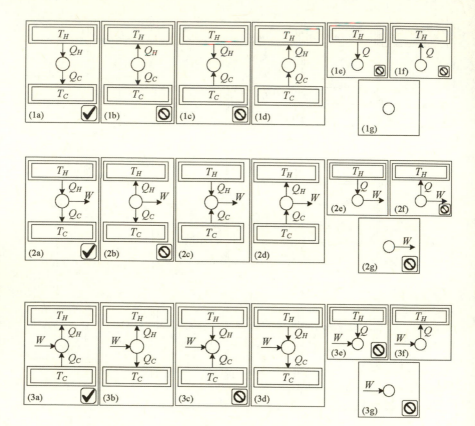

FIGURE B.2 Catalog of 21 simple cycles. Explicitly allowed cycles are marked ✓ and cycles necessarily prohibited by the first law are marked ⊘.

cycle (2e) on a copy of the catalog with a forbidden symbol, ⊘. To find which others of the 21 cycles necessarily violate Thomson's second law requires more effort. As an example, consider the catalog cycle (1e), which extracts heat from a single heat reservoir without producing or consuming work. We can prove in the following way that this cycle necessarily violates both Carnot's and Thomson's second law. The cyclic process in question is denoted 1 in Figure B.3. We combine supposed cycle 1 with the explicitly allowed heat engine cycle 2. The two cycles are adjusted so that supposed cycle

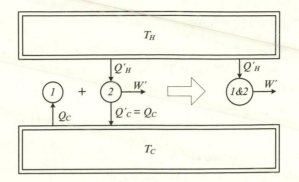

FIGURE B.3 Indirect proof that cyclic process 1 is necessarily forbidden by both Carnot's and Thomson's second laws. Supposing that cyclic process 1 is possible necessarily leads to an engine 1&2 that violates the Carnot and Thomson statements of the second law.

1 extracts the same heat from the reservoir at temperature T_C that the explicitly allowed heat engine 2 rejects to that reservoir. Thus, $Q_C = Q'_C$. The net result is a combined cyclic process 1&2 that simply extracts heat from a single reservoir and produces work W'—a clear violation of both Carnot's and Thomson's second laws. This violation follows for any positive initial values of Q_C, Q'_H, Q'_C, and W'. Therefore, we should mark cycle (1e) of both the Thomson and the Carnot catalogs with the symbol $\oslash$.

Any two logically equivalent versions of the second law should have identical catalogs, that is—have an identical pattern of ✓ and $\oslash$ markings. Likewise, any two catalogs with identical markings must follow from logically equivalent laws. For instance, since Carnot's second law is logically equivalent to Clausius's second law, these two versions of the second law should allow and forbid exactly the same simple cycles. And since Thomson's second law is less restrictive than either Carnot's or Clausius's, the former should forbid fewer cycles than either of the latter.

Appendix B Problems

B.1 *Thomson.* Which of the 21 simple cycles represented in Figure B.1 are necessarily forbidden by Thomson's statement of the second law of thermodynamics?

B.2 *Necessarily Forbidden.* Determine which of the 21 simple cycles are necessarily forbidden
(a) by Carnot's second law;
(b) by Clausius's second law.

B.3 *Another Second Law.* Show that the following version of the second law is at least as restrictive as Thomson's second law by finding which of its catalog cycles are necessarily forbidden:

A cyclic process that extracts heat from at least one reservoir but does not reject heat to another reservoir is impossible.

B.4 *First and Second Laws.* Show that the Thomson and the Carnot/ Clausius statements of the second law necessarily forbid the same catalog cycles when each is assumed in conjunction with the first law.

Appendix C

....

Glossary of Terms

Adiabatic boundary. A boundary that prohibits heat interactions.

Arrow of time. A property of isolated thermodynamic systems that requires their entropy to increase during an irreversible process.

Boundary. The surface that separates a system from its environment. A boundary permits or forbids work to be done on or by the system, permits or forbids heat exchange, and permits or forbids matter to enter or leave the system.

Cycle. A sequence of interactions that returns a system to its initial state.

Diathermal boundary. A boundary that permits heat exchange.

Efficiency, heat engine. The ratio of the work produced by a heat engine to the heat it absorbs from the hotter of the two reservoirs with which it interacts.

Entropy. An extensive thermodynamic state variable that quantifies a system's accumulated irreversible change relative to a reference state.

Equation of state. A relation among state variables.

Equilibrium process. Same as quasistatic process.

Equilibrium state. A state that changes only when the system's environment changes.

Fluid. A class of systems that can be adequately described by a temperature, a volume, and a single-valued pressure.

Heat. That which when transferred to or from a system through an impermeable, work-prohibiting boundary changes the system's state.

Heat capacity. The quantity of heat absorbed or rejected by a system per unit change in temperature caused by that interaction.

Heat pump. A heat engine that transfers heat from a cold body to a hot body in order to maintain the temperature of the hot body.

Heat reservoir. An object to and from which indefinite quantities of heat can be transferred without changing its temperature. The heat capacity of a heat reservoir is indefinitely large.

Ice point, normal. The temperature at which ice melts at atmospheric pressure.

Intensive state variable. A thermodynamic variable that is independent of the size of the system.

Intrinsic stability. The stability of a homogeneous system arising from the mutual stability among its spatially separate parts.

Isentropic process. An entropy-conserving process.

Isobaric process. A constant pressure process.

Isochoric process. A constant volume process.

Isothermal process. A constant temperature process.

Latent heat. The net quantity of heat absorbed or rejected during a phase transition. The absorption or rejection of latent heat leaves a system's temperature unchanged.

Latent heat of fusion. The heat required to melt one gram of solid at its normal melting point.

Latent heat of vaporization. The heat required to vaporize one gram of liquid at its normal boiling point.

Mechanical equivalent of heat. The ratio of work performed on a system to the heat absorbed by the same system that will cause the same increase in temperature. Its value is 4.186 joules/calorie.

Molar specific heat. Heat capacity per unit mole.

Nernst postulate. The third law of thermodynamics, especially those versions originating with the German physical chemist Walther Nernst (1864–1941).

Principle of Thomsen and Berthelot. A rule of thumb stating that chemical reactants realize the most exothermic reaction possible.

Quasistatic process. A series of changes that unfold so slowly that the system always remains arbitrarily close to an equilibrium state.

Refrigerator. A heat engine that transfers heat from a cold body to a hot body in order to maintain the temperature of the cold body.

Reversible process. A process that proceeds indefinitely slowly, that is, quasistatically, and without friction or internal dissipation.

Specific heat. Heat capacity per unit mass.

Stable equilibrium. The equilibrium state of a system that admits no thermodynamic change.

State. A system's condition as determined by the values of its state variables. A thermodynamic state is described by a relatively small number of variables.

State variables. Variables whose value describes a system. The state variables of a system may change when the system interacts with its environment.

Steam point, normal. The temperature at which liquid water vaporizes at atmospheric pressure.

System. That part of the universe with which one is concerned. Each system is surrounded by a boundary with definite properties.

Temperature, empirical. Temperature as determined by the empirical state of a particular system.

Temperature, thermodynamic. Temperature as determined by either the heat ratio or the efficiency of a Carnot cycle.

Thermal equilibrium. The relationship of two systems when those systems are such that if allowed to interact thermally through a work-prohibiting, impermeable boundary do not change their state.

Thermometer. An equilibrium-indicating system.

Triple point. The thermodynamic state at which the solid, liquid, and vapor phases of a pure substance coexist.

Vapor dome. The region of thermodynamic phase space in which a liquid and its vapor coexist.

Appendix D

.....

Selected Worked Problems

1.3 *Interactions*. In each of the following interactions indicate whether the system indicated does work, has work done upon it, or does no work and whether the system boundary is diathermal or adiabatic.

(a) The system is the air contained within a bicycle tire along with a tire pump connected to it. The pump plunger is pushed down, forcing air into the tire. Assume this interaction is over before the air is significantly cooled.

Work is done on the air as the pump handle is forced down. The interaction occurs relatively quickly, and the air does not have time to cool—that is, there is no time for heat to be transferred from the air inside the pump to its environment. Thus, the system boundary is effectively adiabatic.

4.7 *First Law Equivalent*. The first law of thermodynamics can be formulated as the following statement of impossibility: "It is impossible to devise a cycle that has no effect other than the performance of work on or by the environment."

(a) Use an indirect proof to show that this impossibility version of the first law, the existence of heat engine and refrigerator cycles diagrammed in Figure 4.2a–b, and the possibility of adjusting and combining cycles, as explained in Section 4.5, together lead to the denial of the one-flow heat cycles diagrammed in Figure 4.3c–d.

In constructing an indirect proof we assume the contradiction of the statement to be proven and argue to some kind of absurdity. Given the purpose of both parts of this problem, the absurdity with which we wish to conclude is the possibility of the two one-flow heat cycles diagrammed in Figure 4.3c–d. The contradiction of the stated impossibility version of the first law is "It is possible to devise a cycle that has no effect other than the performance of work on or by the environment." Thus, we can devise cycles

184

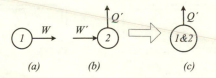

(a)　　　(b)　　　(c)

FIGURE D.1 A supposed cycle (on the left) that simply produces work is combined with a legitimate refrigerator cycle (in the middle) to produce a cycle (on the right) that simply produces heat. The combination is made by adjusting the two cycles on the left until $W = W'$.

that simply produce or consume work. We adjust these cycles so that they supply or consume the work consumed or produced by the legitimate heat engine and refrigerator cycles as diagrammed in Figure 4.2a–b. One such combination of two cycles is shown in Figure D.1.

A second cycle that simply consumes work is constructed in very similar fashion. These two cycles produce the absurdity consisting of declaring that the two one-flow heat cycles diagrammed in Figure 4.3c–d are possible, that is, of contradicting the impossibility of the two one-flow heat cycles.

5.2 *Clausius's Second Law.* Prove that the following processes lead to a violation of Clausius's second law:
(a) A cyclic process that absorbs heat from a reservoir and produces work as shown in Figure 5.2b.

The strategy of this proof is illustrated in Figure D.2. On the left we see the supposed cycle that absorbs heat from a reservoir and produces work. Since this supposed cycle draws heat from only one reservoir, its temperature as well as its placement above or below the cycle symbol is arbitrary. This supposed cycle is combined with the explicitly allowed refrigerator cycle that is also shown. The work produced by the supposed cycle, W, and the work consumed by the refrigerator, W', are adjusted until $W = W'$. The combined cycle merely extracts heat $Q_C + Q'_C$ from the T_C reservoir and rejects heat Q_H to the hotter T_H reservoir—clearly a violation of Clausius's second law.

6.1 *Engine Efficiencies*
(a) What is the maximum efficiency of a heat engine operating between reservoirs with temperatures of 20°C and 500°C?

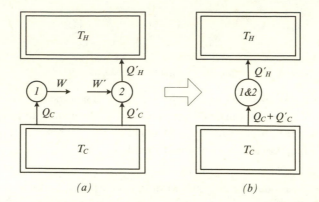

(a)   *(b)*

FIGURE D.2   A supposed cycle (on the left) that extracts heat from a single reservoir and produces work is combined with a legitimate refrigerator cycle (in the middle) to produce a cycle (on the right) that simply extracts heat from a colder reservoir and rejects heat to a hotter one. The combined cycle is made by adjusting the supposed and legitimate cycles until $W = W'$.

The maximum efficiency, $\varepsilon = 1 - T_C/T_H$, is achieved by a Carnot cycle operating between a hotter reservoir with temperature T_H and a colder one with temperature T_C. However, this relation applies only when the temperatures are thermodynamic temperatures. We change Celsius temperatures to Kelvin ones by adding 273° to the former. Then the maximum efficiency of this engine becomes $\varepsilon = 1 - (20° + 273°) / (500° + 273°)$, that is, 62%.

7.2 *Entropy Change, I.* Does the entropy of the following system increase, decrease, or remain the same as it experiences each of the following changes of state?

(a) One gram of water absorbs enough heat at 373 K and atmospheric pressure to evaporate.

The water absorbs heat at a constant 373 K temperature. Since the system absorbs heat, its entropy increases.

7.3 *Entropy Change, II.* Calculate the entropy change of these systems as a result of the following processes. (When necessary, use the data supplied in Tables 3.1 and 3.2.) Express all answers in SI units.

(a) Twenty-five grams of aluminum melts.

From Table 3.2 we find that the heat of fusion of aluminum is 95.3 cal/g and that its melting temperature is 660°C. Thus, 25 g of aluminum absorb 25×95.3 cal as it melts. Since the melting is a constant-temperature process, the change in entropy is given by $\Delta S = Q/T$, that is, $(25 \times 95.3)/(660 + 273) = 2.55 \cdot$ cal/K, or 10.7 J/K.

7.4 *Carnot Cycle.* Recall that a Carnot cycle is a reversible cycle that extracts heat Q_H from a hot reservoir at temperature T_H, produces work W, and rejects waste heat Q_C to a cold reservoir at temperature $T_C < T_H$. . . . (See Chapter 7 for complete problem description.)

(a) The heat absorbed during the isothermal process $(S_1, T_H) \rightarrow (S_2, T_H)$.

Since during an isothermal process $\Delta S = Q/T$, where Q is the heat absorbed, therefore, $Q = T\Delta S$, and in this case $Q = T_H(S_2 - S_1)$.

8.5 *Hypothetical Characterizing Function.* Suppose the characterizing function of a fluid takes the form

$$E(S,V) = a\frac{S^2}{V},$$

where a is a constant that characterizes the particular hypothetical system.

(a) Determine the equations of state that assume the form $T = T(S,V)$ and $P = P(S,V)$.

We first need to remind ourselves of the fundamental constraint for which the energy, E, is a dependent variable and S and V are independent variables: $dE = TdS - PdV$. From this form we see that $T = (\partial E/\partial S)_V$. Now ours is the straightforward task of taking the derivative of the function of E given above with respect to S while holding V constant and setting the result equal to T. The result: $T = 2aS/V$. A similar procedure produces $P = aS^2/V^2$.

8.17 *Joule-Thomson Coefficient.* Start with the fundamental constraint $dH = TdS + VdP$, form each term into a derivative with respect to P with T held constant, and use one of the Maxwell relations to derive an expression for the so-called *Joule-Thomson coefficient* $(\partial T/\partial P)_H = (V/C_P)$ $(T\alpha_p - 1)$ where $C_P = (\partial H/\partial T)_P$ and the isobaric expansivity $\alpha_p \equiv V^{-1}(\partial V/\partial T)_P$. (See Chapter 8 problems for a complete problem statement.)

Starting with $dH = TdS + VdP$ we form, as directed,

$$\left(\frac{\partial H}{\partial P}\right)_T = T\left(\frac{\partial S}{\partial P}\right)_T + V.$$

First, apply the reciprocity relation involving the three variables, H, P, and T, that is,

$$\left(\frac{\partial H}{\partial P}\right)_T \left(\frac{\partial P}{\partial T}\right)_H \left(\frac{\partial T}{\partial H}\right)_P = -1,$$

to find that

$$\left(\frac{\partial H}{\partial P}\right)_T = -\left(\frac{\partial H}{\partial T}\right)_P \left(\frac{\partial T}{\partial P}\right)_H.$$

Use this result on the left-hand side and recognize that $C_P = (\partial H/\partial T)_P$. Thus,

$$-C_P\left(\frac{\partial T}{\partial P}\right)_H = T\left(\frac{\partial S}{\partial P}\right)_T + V.$$

The derivative $(\partial S/\partial P)_T$ can be transformed with the Maxwell relation $(\partial S/\partial P)_T = -(\partial V/\partial T)_P$ the right-hand side of which helps compose the coefficient $\alpha_p = V^{-1}(\partial V/\partial T)_P$. The desired result follows immediately.

9.3 *Adiabatic Transformation.* A reversible adiabatic transformation is one for which $dQ_{rev} = 0$. Starting from the first law of thermodynamics for reversible, adiabatic transformations of an ideal gas show that
   (a) During a reversible adiabatic transformation of an ideal gas the following are constant: $TV^{\gamma-1}$, PV^γ, and $P^{1-\gamma}T^\gamma$. Here $\gamma \equiv C_P/C_V$.

The first law of thermodynamics for a reversible transformation of a fluid is $dE = dQ - PdV$. When the transformation is also adiabatic, $dQ = 0$ and thus $dE = -PdV$. When the gas is ideal with constant heat capacity C_V, $P = nRT/V$ and $E = C_V T$. Making these substitutions into the differential equation $dE = -PdV$ produces $C_V dT = -(nRT/V)dV$, that is,

$$\frac{dT}{T} = -\left(\frac{nR}{C_V}\right)\frac{dV}{V},$$

which is integrated to produce $\ln[TV^{nR/C_V}] = constant$, that is, $TV^{nR/C_V} = constant$. The ideal gas identity $C_P - C_V = nR$ and the definition $\gamma \equiv C_P/C_V$ turn this result into the first of the requested adiabatic constants.

The other constants follow from using the equation of state $PV = nRT$ to eliminate undesired variables in favor of desired ones. For instance, using $V = nRT/P$ to eliminate V in $TV^{\gamma-1} = constant$ produces $T(nRT/P)^{\gamma-1} = T^\gamma P^{1-\gamma}(nR)^{\gamma-1} = constant$. This simplifies to $T^\gamma P^{1-\gamma} = constant$, where we have changed the meaning of the constant term. Finally, the third adiabatic constant, PV^γ, is derived by using the ideal gas equation of state to eliminate T from $TV^{\gamma-1} = constant$.

9.10 *Room-Temperature Solid.* The energy-characterizing function for a room-temperature solid is

$$E(S,V) = \frac{(V - V_o)^2}{2\kappa_{To} V_o} + E_o \exp\left\{\frac{S}{C_V} - \frac{\alpha_{Po} V}{C_V \kappa_{To}}\right\},$$

where E_o is an arbitrary integration constant with energy units.

(a) Show that the two equations of state for a room-temperature solid,
$P = \alpha_{Po} T/\kappa_{To} + (1 - V/V_o)/\kappa_{To}$ and $E = C_V T + (V - V_o)^2/(2\kappa_{To} V_o)$, can be generated from $E(S,V)$ and appropriate derivatives of $E(S,V)$.

The fundamental constraint $dE = TdS - PdV$ alerts us to the fact that the associated equations of state take the form $T = (\partial E/\partial S)_V$ and $P = -(\partial E/\partial V)_S$. Taking the partial derivative of the given function $E(S,V)$ with respect to S generates

$$T = \left(\frac{E_o}{C_V}\right)\exp\left\{\frac{S}{C_V} - \frac{\alpha_{Po} V}{C_V \kappa_{To}}\right\}.$$

This is a legitimate equation of state but is not in requested form. The requested form follows from using the given function $E(S,V)$ to eliminate the entropy S from the above expression in favor of E and V. In practice we need only solve the above for the exponential term, that is,

$$\exp\left\{\frac{S}{C_V} - \frac{\alpha_{P_0}V}{C_V\kappa_{T_0}}\right\} = \frac{C_V T}{E_o},$$

and eliminate this term from the functional form $E(S,V)$. The equation of state following directly from $P = -(\partial E/\partial V)_S$ must be similarly transformed by eliminating the exponential factor in order to produce the requested form.

9.12 Cavity Radiation, II

(a) Show that the characterizing function $E(S,V)$ for cavity radiation is given by $E(S,V) = (3S/4)^{4/3}/(aV)^{1/3}$.

In order to derive the characterizing function, $E(S,V)$, for cavity radiation, we start from its two equations of state: $P = E/3V$ and $E = aVT^4$. Then we produce differential equations which can be integrated to yield either $E(S,V)$ or $S(E,V)$. If the latter is the result, then we solve for the former. The fundamental relation $dE = TdS - PdV$ produces $T = (\partial E/\partial S)_V$ and $P = -(\partial E/\partial V)_S$, while $dS = dE/T + (P/T)dV$ leads to $1/T = (\partial S/\partial E)_V$ and $P/T = (\partial S/\partial V)_E$. The integrations are more straightforward when we use the latter two equations of state. We use the given equations of state to eliminate all but the independent variables E and V from the left-hand side of the two formal equations of state. Thus,

$$\left(\frac{\partial S}{\partial E}\right)_V = \left(\frac{aV}{E}\right)^{1/4}$$

and

$$\left(\frac{\partial S}{\partial V}\right)_E = \frac{a^{1/4}E^{3/4}}{3V^{3/4}}.$$

In integrating these differential equations it is important to account for the partial derivatives. For instance, when we integrate the first by holding V constant, any integration constants will be arbitrary functions of V. Thus,

$dS = (aV)^{1/4}E^{-1/4}dE$ integrates to $S(E,V) = 4(aV)^{1/4}E^{3/4}/3 + f(V)$, where the function $f(V)$ is as yet undetermined. In similar fashion the second differential equation integrates to $S(E,V) = 4a^{1/4}E^{3/4}V^{1/4}/3 + g(E)$. These two expressions for $S(E,V)$ are compatible if and only if $f(V)$ and $g(V)$ are identically zero. Then, $S(E,V) = 4(aV)^{1/4}E^{3/4}/3$, which when inverted produces the desired expression, $E(S,V) = (3S/4)^{4/3}/(aV)^{1/3}$.

10.2 *Paramagnetic Material.* Show that the energy, E, of a paramagnetic material that obeys Curie's law is a function of temperature, T, alone. Work only from what we know about paramagnetic materials.

What we know about paramagnetic materials includes the fundamental constraint $dE = TdS + (B_o/\mu_o)dB_m$ and Curie's law, $B_m = \mathcal{C}B_o/T$. Since we want to derive a result on the relationship of the energy, E, to the variables B_m, B_o, and T, we seek formal relations that contain only these variables and not the entropy. Such a relation is found by recasting the fundamental relation in entropy form, that is, $dS = dE/T - (B_o/\mu_o T)dB_m$. From this we extract the cross-differentiation

$$\frac{\partial}{\partial B_m}\left(\frac{1}{T}\right)_E = -\frac{\partial}{\partial E}\left(\frac{B_o}{\mu_o T}\right)_{B_m}.$$

Making use of Curie's law to eliminate B_o from the right-hand side produces

$$\frac{\partial}{\partial B_m}\left(\frac{1}{T}\right)_E = -\frac{\partial}{\partial E}\left(\frac{B_m}{\mu_o \mathcal{C}}\right)_{B_m}.$$

The right-hand side vanishes and the left-hand side becomes $(\partial T/\partial B_m)_E = 0$. Thus, the temperature, T, is not a function of B_m but only of energy, E, that is, $T = T(E)$. The inverse of this relationship shows that energy is a function of temperature alone.

10.6 *Eötvös Equation.* The Eötvös equation of state for surface tension is $\sigma(T) = \sigma_o(1 - T/T_c)$, where $T \leq T_c$. Show that, given the Eötvös equation of state and the context of the theoretical picture of Section 10.4,
   (a) the internal energy $E = \sigma_o A$, and
   (b) the entropy $S = \sigma_o A/T_c$.

The fundamental constraint for a surface is $dE = TdS + \sigma dA$, where from what we know of the general theoretical context of section 10.4 $E(T) = [\sigma(T) - T(d\sigma/dT)]A$. Substituting the Eötvös equation of state into this expression produces the answer given in part (a). To find an expression for the entropy we write the fundamental constraint in entropy form, that is, $dS = dE/T - (\sigma/T)dA$. Since $E = \sigma_o A$, substitution into this version of the fundamental constraint yields $dS = (\sigma_o - \sigma)dA/T$, which, given the Eötvös equation, reduces to $dS = \sigma_o dA/T_c$. The latter can be integrated directly to produce the result sought in part (b).

10.12 Prove the following relations among variables describing a single-phase, open fluid system.

(a) $\left(\dfrac{\partial T}{\partial n}\right)_{S,P} = \left(\dfrac{\partial \mu}{\partial S}\right)_{P,n}$ and $\left(\dfrac{\partial V}{\partial n}\right)_{S,P} = \left(\dfrac{\partial \mu}{\partial P}\right)_{S,n}$.

The key to deriving these relations is to start with the correct fundamental constraint and make formal deductions from it. The independent variables appearing in the first relationship above, $(\partial T/\partial n)_{S,P} = (\partial \mu/\partial S)_{P,n}$, are S, P, and n. We need a version of the fundamental constraint for an open fluid system with these independent variables. Transforming the characterizing function from energy to enthalpy, $H(= E + PV)$, produces the desired form, $dH = TdS + VdP + \mu dn$. The equations of state that follow from this form are $T = (\partial H/\partial S)_{P,n}$, $V = (\partial H/\partial P)_{S,n}$, and $\mu = (\partial H/\partial n)_{S,P}$. The relations sought follow from equating the appropriate cross-differentiations. For instance, since $\partial^2 H/\partial n\partial S = \partial^2 H/\partial S\partial n$, then $(\partial T/\partial n)_{S,P} = (\partial \mu/\partial S)_{P,n}$. The other Maxwell-like relations follow in similar fashion from other cross-differentiations.

12.7 *Saturated Vapor Model.* Assume that the volume of a saturated liquid is ignorably small compared with the volume of the saturated vapor at the same temperature and that the ideal gas equation of state, $PV = nRT$, describes the saturated vapor.

(a) Given these assumptions derive an expression for $P(T)$ within a region of the vapor dome for which the heat of transition $Q_{l \to v}$ is a constant independent of temperature. Use the initial condition $P = P_o$ when $T = T_o$ to evaluate the integration constant. [Hint: Use Equation 12.31.]

We start with the Clausius-Clapeyron equation, $dP/dT = Q_{a\to b}/T(V_b - V_a)$, since this regulates the function $P(T)$ within a phase transition. Let state b be that of the saturated vapor and state a be that of the saturated liquid. Then $Q_{a\to b}$ is the latent heat of vaporization. Given that the volume of the saturated liquid, V_a, is ignorably small compared with the volume of the saturated vapor, V_b, at the same temperature, $dP/dT = Q_{a\to b}/TV_b$. Furthermore, we treat the saturated vapor as an ideal gas so that $V_b = nRT/P$. Thus, we find Equation 12.31, $dP/dT = PQ_{a\to b}/nRT^2$. Since P and T are the only variables, we may separate these variables and integrate the result using the given initial conditions. Thus,

$$\frac{dP}{P} = \frac{Q_{a\to b}}{nR} \frac{dT}{T^2}$$

so that

$$\int_{P_0}^{P} \frac{dP}{P} = \frac{Q_{a\to b}}{nR} \int_{T_0}^{T} \frac{dT}{T^2}.$$

Completing the integration produces $\ln(P/P_0) = (Q_{a\to b}/nR)(1/T_0 - 1/T)$, which can be solved for the desired function, $P(T) = P_0 \exp[(Q_{a\to b}/nR)(1/T_0 - 1/T)]$.

13.1 *Low-Temperature Limit of Cavity Radiation.* Many of the equations of state discussed in this volume incorporate a constant heat capacity C_V appropriate only for high-temperature systems. However, the equations of state for cavity radiation, in Section 9.3, should obtain for all temperatures, including those down to the $T \to 0$ limit. Show that cavity radiation observes the following limits.

(a) $\left(\dfrac{\partial S}{\partial V}\right)_T \to 0$ as $T \to 0$.

First the function $S(V,T)$ for cavity radiation must be found. Then the partial derivative will follow easily. The quickest route is to start with the characterizing function $E(S,V) = (3S/4)^{4/3}/(aV)^{1/3}$ given in Problem 9.12, solve for $S(V,E)$, and then eliminate E in favor of T and V via the equation of state $E = aVT^4$. Accordingly, $S(T,V) = 4aVT^3/3$. Thus, $(\partial S/\partial V)_T = 4aT^3/3$, which vanishes in the $T \to 0$ limit.

Appendix E

....

Answers to Problems

1.2 Work

1.3 (a) Adiabatic; work done on system. (b) Diathermal; no work done. (c) Adiabatic; system does work. (d) Diathermal; work done on system. (e) Adiabatic; no work done.

2.1 (a) $R_o = 7.000\ \Omega$, a = $3.92\ 10^{-3}/°C$, $b = -5.82\ 10^{-7}/°C^2$.

(b) $T = \left(a/2b\right)\left[\sqrt{1 + \left(4b/a^2\right)\left(R/R_o - 1\right)} - 1\right]$.

(c) 421.1°C.

2.2 $-40°C$ and $-40°F$

2.3 26 Ré

3.1 28 cal

3.2 0.221 cal/g °C

3.3 30.8°C or 87.4°F

3.4 1500 cal

4.1 5.48 J/cal

4.2 0.571°C

4.3 -11.6 J

4.4 (a) No. (b) Yes. (c) Positive.

4.5 Internal energy remains the same

4.6 (a) 0.172 J. (b) -0.122 J.

4.8 527 J of work is done by the system

6.1 (a) 62%. (b) 40%.

6.2 0.043 cal

6.3 (a) 13. (b) 20.

6.4 (a) $(1 - T_H/T_C)/(1 + a) + (1 - T_M/T_C)a/(1 + a)$

6.6 (a) $\dot{W} = \sigma_B A T_C^4 \left(\dfrac{T_H}{T_C} - 1 \right)$.

(b) $A = \dfrac{4^4 \dot{W}}{3^3 \sigma_B T_H^4}$ and $T_C = \dfrac{3}{4} T_H$.

6.7 $Q_3/T_3 + Q_2/T_2 - Q_1/T_1 \leq 0$

6.8 (c) $W = C\left(T_H + T_C - 2\sqrt{T_H T_C} \right)$

7.1 $C \ln(T_f/T_i)$

7.2 (a) Increases. (b) Decreases. (c) Increases. (d) Remains the same.
(e) Decreases. (f) Increases. (g) Increases.

7.3 (a) 10.7 J/K. (b) –60.5 J/K. (c) 14.1 J/K.

7.4 (a) $T_H(S_2 - S_1)$. (b) $T_C(S_1 - S_2)$. (c) $(T_H - T_C)(S_2 - S_1)$.
(d) $(T_H - T_C)(S_2 - S_1)$.

7.5 (a) No change. (b) $+8.05 \times 10^{-3}$ J/K.

7.6 (a) $C \ln[(T_H + T_C)^2/4T_H T_C]$. (b) Zero.

7.7 (a) –35.6 J/K. (b) 36.3 J/K.

8.1 *Left to right:* Row (a): 0.49 atm, 380 torr, 380 mm Hg.

Row (b): 1.01×10^5 Pa, 760 torr, 760 mm Hg.

Row (c): 1.30×10^{-4} Pa, 1.30×10^{-9} atm, 10^{-6} mm Hg.

Row (d): 4.0×10^3 Pa, 4.0×10^{-2} atm, 30 torr.

8.2 $E = bPV^2/n$, where $b = 1.25 \times 10^{-4}$ mole/m^3

8.3 (a) Path ABC. (b) Path ABC. (c) 2×10^5 J, 2×10^5 J. (d) 0. (e) 0.

8.4 4.7 atm

8.5   (a) $T = 2aS/V$, $P = aS^2/V^2$.

8.6   (a) $1/T = (\partial S/\partial E)_V$, $P/T = (\partial S/\partial V)_E$.
      (b) $(\partial/\partial V)_E(1/T) = (\partial/\partial E)_V(P/T)$.

8.7   (a) 26,000 cal = 1.09×10^5 J. (b) 3.38×10^4 J.

8.12   (a) $P = -(\partial A/\partial V)_T$. (b) $S = -(\partial A/\partial T)_V$. (c) $E = A - T(\partial A/\partial T)_V$.
      (d) $C_V = -T(\partial^2 A/\partial T^2)_V$. (e) $\kappa_T = 1/[V(\partial^2 A/\partial V^2)_T]$.
      (f) $\alpha_P = -(\partial^2 A/\partial V \partial T)/V(\partial^2 A/\partial V^2)_T$.
      (g) $\alpha_V = (\partial^2 A/\partial T \partial V)/(\partial A/\partial V)_T$.

8.15   Extensive: V, E, S, PV, H, C_V, n, and $(\partial V/\partial T)_P$. Intensive: P, T,
      and c_V. Neither: E^2, T/H, and κ_T.

9.4   (a) $T^{C-C_V}V^{C_V-C_P} = const.$

9.5   3.17 atm

9.6   44.3 atm

9.7   (a) $W_{1\to2} = nRT_H \ln(V_2/V_1)$. (b) $Q_H = nRT_H \ln(V_2/V_1)$.
      (c) $W_{2\to3} = C_V(T_H - T_C)$.
      (d) $W = nRT_H \ln(V_2/V_1) + nRT_C \ln(V_4/V_3)$.

9.8   (a) $\Delta E = 0$. (b) $\Delta S = nR\ln(V_f/V_i)$. (c) $P_f = P_iV_i/V_f$.
      (d) $T_f = P_iV_i/nR$.

9.9   (a) $nR(T_2 - T_1)/(\gamma - 1)$. (b) $C_V(T_3 - T_2)$. (c) $nR(T_3 - T_4)/(\gamma - 1)$.
      (d) $C_V(T_4 - T_1)$. (e) $T_3 > T_4 > T_2 > T_1$. (f) $1 - r^{1-\gamma}$. (g) 0.602.

9.14   (a) $W_{1\to2} = (aT_H^4/3)(V_2 - V_1)$. (b) $Q_H = (4aT_H^4/3)(V_2 - V_1)$.
      (c) $W_{2\to3} = a(V_2T_H^4 - V_3T_C^4)$.
      (d) $W = (aT_H^4/3)(V_2 - V_1) + a(V_2T_H^4 - V_3T_C^4) + (aT_C^4/3)(V_4 - V_3)$
           $+ a(V_4T_C^4 - V_1T_H^4)$.

9.15   (a) Ideal gas pressure: 1.02×10^{11} atm; radiation pressure: 1.27
      $\times 10^8$ atm; ratio = 809. (b) Ideal gas pressure: 8.11×10^{-2} atm;
      radiation pressure: 2.82×10^{-6} atm; ratio = 2.88×10^4.

9.16   1.38 kW/m^2

10.1 $(\partial T/\partial L)_S = (\partial F/\partial S)_L; -(\partial S/\partial L)_T = (\partial F/\partial T)_L; (\partial T/\partial F)_S = -(\partial L/\partial S)_F;$
and $(\partial S/\partial F)_T = (\partial L/\partial T)_F$

10.3 $\mu_o(\partial T/\partial B_m)_S = (\partial B_o/\partial S)_{B_m}; -\mu_o(\partial S/\partial B_m)_T = (\partial B_o/\partial T)_{B_m};$
$-\mu_o(\partial T/\partial B_0)_S = (\partial B_m/\partial S)_{B_o}; $ and $\mu_o(\partial S/\partial B_o)_T = (\partial B_m/\partial T)_{B_o}$

10.4 2.87×10^{-2} J

10.7 (a) $E(A,T) = \sigma_o A\left(1 - \dfrac{T}{T_c}\right)^{n-1}\left[1 + \dfrac{T}{T_c}(n-1)\right].$

(b) $S(A,T) = \dfrac{An\sigma_o}{T_c}\left(1 - \dfrac{T}{T_c}\right)^{n-1}.$

11.1 The volume V; it is minimized.

12.3 8.45×10^3 cm^3

12.5 141°C

12.7 (a) $P/P_o = \exp\{(Q_{l\to v}/nR)(T_o^{-1} - T^{-1})\}.$ (b) 28°C (50°F).

12.9 $P_c = (a'/4b'^2)e^{-2}, V_c = 2nb', T_c = a'/4Rb',$ and
$P_c V_c/nRT_c = 2e^{-2} = 0.271$

12.10 $P_c = B_o\left(\dfrac{m}{n+3}\right)^{(n+3)/(n-m)}\left(\dfrac{m+3}{n}\right)^{(m+3)/(n-m)},$

$V_c = V_o\left(\dfrac{n+3}{m}\right)^{3/(n-m)}\left(\dfrac{n}{m+3}\right)^{3/(n-m)},$

$T_c = \dfrac{B_o V_o}{n_m R}\left(\dfrac{m}{n+3}\right)^{m/(n-m)}\left(\dfrac{m+3}{n}\right)^{n/(n-m)},$

$\dfrac{P_c V_c}{n_m R T_c} = \dfrac{nm}{(n+3)(m+3)}$

B.1 (1c), (1e), (2c), and (2e)

B.2 Both (a) and (b): (1b)–(1f), (2b)–(2f)

B.3 (1c), (1e), (2c), (2e), (3c), (3e), and (3g)

Annotated Bibliography

An item appears here for at least one, and usually for both, of two reasons: I consulted it in writing the text, and/or I recommend it for further study. Publication dates always refer to the first edition unless two dates appear. Then the earlier date refers to the edition actually consulted and the second one (in parentheses) to the first edition.

Adkins, C. J. *Equilibrium Thermodynamics*. 2nd ed. London: McGraw-Hill, 1975 (1968).

A carefully written textbook, more complete and at a higher level than *Mere Thermodynamics*. Contains a description of C. Caratheodory's reformulation of thermodynamics. 284 pages.

Baierlein, Ralph. *Thermal Physics*. Cambridge: Cambridge University Press, 1999.

This text and the one by Daniel Schroeder (listed below) are good examples of the "thermal physics" approach that integrates classical thermodynamics and statistical physics into one narrative. 442 pages.

Brown, Sanborn C. *Count Rumford—Physicist Extraordinary*. New York: Anchor, 1962.

A short, popular biography of this colorful figure by a professional physicist. Quotes from original sources. 178 pages.

Callen, H. B. *Thermodynamics*. New York: John Wiley & Sons, 1960.

A pioneering text that develops the subject from a set of simple, independent postulates. Callen's approach falls between those that proceed from the first and second laws of thermodynamics and those, like Caratheodory's, that proceed from minimal assumptions. 376 pages.

Carnot, Sadi. *Reflections on the Motive Power of Fire*. Introduced and translated by E. Mendoza. New York: Dover, 1960.

Contains seminal papers by S. Carnot (1824), E. Clapeyron (1834), and R. Clausius (1850). The first 12 pages of Carnot's *Reflections* lucidly present the foundations of the second law of thermodynamics without presuming either the first law or the special properties of ideal gases. For those with a taste for original sources. 152 pages.

The Dictionary of Scientific Biography. Edited by Charles Coulston Gillespie. New York: Scribner's, 1974.

The standard reliable source in 18 volumes.

Eddington, Arthur. *The Nature of the Physical World*. New York: Macmillan, 1928.

Chapter 5 contains a discussion of the thermodynamic basis of what Eddington calls *the arrow of time*. 361 pages.

Einstein, A. "Autobiographical Notes." In *Albert Einstein: Philosopher-Scientist*, ed. P. A. Schilpp. New York: Harper and Row, 1959.

Einstein's assessment of thermodynamics appears on page 33. 600 pages.

Fermi, Enrico. *Thermodynamics*. New York: Dover, 1956 (1936).

A marvel of concision that has withstood the test of time. However, Fermi overuses the ideal gas. 160 pages.

Finn, C. P. B. *Thermal Physics*. 2nd ed. London: Chapman and Hall, 1993 (1986).

In spite of its title, this undergraduate textbook focuses on classical thermodynamics. At about the same level as *Mere Thermodynamics*. 256 pages.

Lemons, Don S., and Carl M. Lund. "Thermodynamics of High Temperature, Mie-Grüneisen Solids." *American Journal of Physics* 67 (1999): 1105–8.

Presents an alternative two-phase equation of state.

Lemons, Don S., and Margaret K. Penner. "Sadi Carnot's Contribution to the Second Law of Thermodynamics." *American Journal of Physics* 76 (2008): 21–25.

Contains a proof that, without assuming the first law, Thomson's and Clausius's versions of the second law are not logically equivalent.

Lewis, Gilbert N., and Merle Randall. *Thermodynamics*. Revised by K. S. Pitzer and L. Brewer. 2nd ed. New York: McGraw-Hall, 1961 (1923).

Early-twentieth-century expansive "introduction to research" and "guide for anyone who wishes to use thermodynamics in productive work." 723 pages.

Magie, W. F. *Source Book in Physics*. Cambridge: Harvard University Press, 1965 (1935).

Excerpts from a large number of important primary sources with biographical introductions. 620 pages.

Mahan, Bruce H. *Elementary Chemical Thermodynamics*. Mineola, N.Y.: Dover, 2006 (1963).

The title says it all. 155 pages.

Mott-Smith, Morton. *The Concept of Energy Simply Explained*. New York: Dover, 1964 (1934).

Good histories of classical thermodynamics are hard to find, but this engaging, if poorly titled, popularization serves the purpose quite well. Frequently quotes original sources. 215 pages.

———. *The Concept of Heat and Its Workings Simply Explained*. New York: Dover 1962 (1933).

Elementary and verbal, yet accurate, description of basic thermodynamic phenomena. 165 pages.

Pippard, Brian. *Elements of Classical Thermodynamics for Advanced Students of Physics*. Cambridge: Cambridge University Press, 1960 (1957).

A precise if somewhat abstract formulation of classical thermodynamics. 165 pages.

Planck, Max. *Treatise of Thermodynamics*. Mineola, N.Y.: Dover, 1990 (1922).

The last chapter contains Planck's version of the third law of thermodynamics. 247 pages.

Potter, Merle C., and Craig W. Somerton. *Engineering Thermodynamics*. New York: McGraw-Hill, 1996 (1993).

Outlines rather than narrates thermodynamics. Full of worked examples and problems arising from engineering practice. 377 pages.

Schamp, H. "Independence of the First and Second Laws of Thermodynamics." *American Journal of Physics* 30 (1962): 825–29.

Argues that one can and should develop the consequences of the second law without assuming the first law.

Schroeder, Daniel. *Thermal Physics*. New York: Addison-Wesley, 1999.

Sklar, Lawrence. *Physics of Chance*. Cambridge: Cambridge University Press, 1993.

Philosophical analysis of statistical mechanics with an informative chapter entitled "The Reduction of Thermodynamics to Statistical Mechanics." 437 pages.

Tabor, D. *Gases, Liquids, and Solids.* 2nd ed. Cambridge: Cambridge University Press, 1979 (1969).

Has very much the same purpose and style as A. J. Walton's *Three Phases of Matter* (listed below) at half the length. 233 pages.

Vanderslice, J. T., H. W. Schamp, Jr., and E. A. Mason. *Thermodynamics.* Englewood Cliffs, N.J.: Prentice-Hall, 1966.

Proof that good books don't always stay in print. This text inspired the tactic, adopted in Chapter 5 and in Appendix B, of considering the consequences of the second law without assuming the first law. At a slightly more advanced level than *Mere Thermodynamics*. 244 pages.

Walton, Alan J. *Three Phases of Matter.* Oxford: Oxford University Press, 1983 (1976).

Designed to "give students a real feel for what solids, liquids, and gases are like at the atomic level." Relates macroscopic properties to microscopic forces. Filled with interesting data and discussions of experimental methods. Complements *Mere Thermodynamics*. 482 pages.

Index

....

The letter *g* following a page number indicates a glossary entry; the letter *p*, a problem; and the letter *w*, a worked problem.